INTELLIGENT DIGITAL OIL AND GAS FIELDS

INTELLIGENT DIGITAL OIL AND GAS FIELDS

Concepts, Collaboration, and Right-Time Decisions

GUSTAVO CARVAJAL
MARKO MAUCEC
STAN CULLICK

Gulf Professional Publishing
An imprint of Elsevier

Gulf Professional Publishing is an imprint of Elsevier
50 Hampshire Street, 5th Floor, Cambridge, MA 02139, United States
The Boulevard, Langford Lane, Kidlington, Oxford, OX5 1GB, United Kingdom

Notices
Knowledge and best practice in this field are constantly changing. As new research and experience
broaden our understanding, changes in research methods, professional practices, or medical treatment
may become necessary.

Practitioners and researchers must always rely on their own experience and knowledge in evaluating
and using any information, methods, compounds, or experiments described herein. In using such
information or methods they should be mindful of their own safety and the safety of others,
including parties for whom they have a professional responsibility.

To the fullest extent of the law, neither the Publisher nor the authors, contributors, or editors,
assume any liability for any injury and/or damage to persons or property as a matter of products
liability, negligence or otherwise, or from any use or operation of any methods, products, instructions,
or ideas contained in the material herein.

Library of Congress Cataloging-in-Publication Data
A catalog record for this book is available from the Library of Congress

British Library Cataloguing-in-Publication Data
A catalogue record for this book is available from the British Library

ISBN: 978-0-12-804642-5

For information on all Gulf Professional publications
visit our website at https://www.elsevier.com/books-and-journals

  Working together
to grow libraries in
developing countries

www.elsevier.com • www.bookaid.org

Publisher: Joe Hayton
Acquisition Editor: Katie Hammon
Editorial Project Manager: Kattie Washington
Production Project Manager: Anitha Sivaraj
Designer: Matthew Limbert

Typeset by SPi Global, India

DEDICATION

Gustavo Carvajal... "To Geraldine, loving you is a wonderful world."

Marko Maucec... "To my loving family, Branka, Katja, Jan, and Jernej, for inspiration, devotion, and unconditional support, no matter how far apart life takes us."

Stan Cullick... "To my wife Pat Dashiell who stood with me over a three-decade career journey with long work hours and many weeks at a time away from home."

CONTENTS

PREFACE

Digitalization of oil and gas operations started several decades ago using supervisory control and data acquisition (SCADA) systems to gather data on high-cost producing assets, such as offshore, and to replace electromechanical paper chart recorders. In the last decade—driven by some key technology trends, such as reduced costs for sensors and data storage (Big Data) and ubiquitous communication—the industry is seeing more implementations of what we all know as the digital oil field (DOF). In general, DOF refers to acquiring high-frequency data, even streaming real-time data, and using that data to increase production efficiency, reduce downtime, implement field-wide efficiencies, and optimize reservoir recovery and management.

While DOF implementations are increasing, as of 2017, the oil and gas (O&G) industry still faces many challenges to achieving full implementation of and maximum value from DOF solutions. Many different industries—such as mobile communications, medical, computers, social media, retail, and industries, such as airlines—report that more than 70% of previously manual processes are now automated and connected in real time to data-driven solutions or the Internet of things (IoT), leading to faster responses and increased operational efficiency. But surveys and news articles from the O&G industry suggest that less than 25% of all firms surveyed are connected to real-time operational data, and the percentage of operating wells that have digital real-time data is less than 10%.

Early on, many O&G firms thought that the DOF was simply "IT" (information technology) or "data management." While we now know that DOF is so much more, the low numbers from these survey results led us to believe that many companies are still struggling to figure out how to implement DOF solutions and realize the promise of its value to operations and business.

We have been fortunate to work on some ground-breaking DOF projects that have delivered on the operational and business performance results. These projects leveraged technology and related cost breakthroughs in areas such as sensors, wireless connectivity, storage, computations, data analytics, process engineering, and automation and control.

We were motivated to write this book to share our experiences in implementing such solutions, and in identifying and overcoming the many challenges faced when implementing DOF solutions in inherently complex O&G operations.

Our work on and knowledge of the DOF began in the mid-2000s. In 2010–13, as part of an extensive team from Halliburton, we worked on a multi-vendor project for Kuwait Oil Company (KOC) called the Kuwait Intelligent Digital Oil Field (KwIDF). The KwIDF project developed state-of-the-industry, real-time data acquisition, and automated production workflows to KOC. KOC and key members of the project team documented the project and its accomplishments by publishing more than 15 technical papers through the Society of Petroleum Engineers (SPE); in 2013, the KwIDF project won the ADIPEC (Abu Dhabi International Petroleum Exhibition and Conference) best-technology award. We acknowledge Kuwait Oil Company's management leadership on KwIDF.

Since our KwIDF experience, we have continued to contribute to, enhance, and develop DOF solutions in our respective positions for companies and projects around the world, including several Middle East fields and in North America in the Permian, Unita, San Joaquin, mid-Continent, and Eagle Ford areas. These projects contributed real value to their respective operators.

Intelligent Digital Oil and Gas Fields: Concepts, Collaboration, and Right-time Decisions draws on our collective and diverse experience to deliver to readers a roadmap through key areas and related issues and challenges of DOF implementations. Chapter 1 sets the stage by providing historical context for details discussed in later chapters, and ending with a summary of some milestone DOF projects and their respective achievements in asset value and financial indicators, serving as incentive to overcome the technical complexities of DOF systems and reach the value.

The book introduces the new age of digital O&G technology and process components, discussing new sensors (SCADA), well mechanics (such as downhole valves), data analytics and models for dealing with a barrage of data, and changes in the way professionals collaborate and make decisions. The book covers topics such as downhole sensors, artificial lift, production surveillance, production optimization, automation, smart wells, integrated reservoir management, and collaboration and change management processes.

The book takes readers on a journey starting at the well level, that is, through instrumentation and measurement for real-time data acquisition, and then provides practical information on analytics on the real-time data. Artificial intelligence techniques provide insights from the data. We discuss workflows on well production decisions from the perspective of "smart," automated and model-based decisions for optimizing production. The journey then moves from the well to the "integrated asset" by detailing how companies use integrated asset models to manage assets (reservoirs) within a DOF context. From model to practice, new ways to operate smart wells enable optimizing an asset. The book also describes collaborative systems and ways of working and how companies are transitioning their work forces to use the technology to make more optimal decisions.

We provide many examples and lessons learned from various case studies, which create a reference that can help managers, engineers, operations, and IT professionals understand specifics on how to filter data, address analytics, and link workflows across the production value chain—that is, its people, processes, and technologies—thereby enabling teams to make better decisions with more certainty and reduced risk.

The DOF started as a small and expensive technology investment. But as the broader information and communication technologies have emerged and costs have plummeted, the O&G industry and DOF systems are leveraging those advances—such as Big Data, ubiquitous communication and lowered costs of DOF sensors and technologies—for significant value. Today large and small O&G companies can use this quickly evolving technology to achieve real value, and it may be more important now than ever. With data volumes and complexity increasing and operating environments and budgets becoming more challenging, engineers and managers are pressed to keep up with infrastructures that are more complex and turn their data into optimized operations and profitable decisions.

This book provides extensive references for further reading and a final chapter on the next-generation DOF. Use it as a reference to help transform engineering workflows and data analytics into successfully delivered O&G projects.

Gustavo Carvajal
Sr. Reservoir Engineer at BP America

Marko Maucec
Petroleum Engineering Specialist at Saudi Aramco

Stan Cullick
Engineering Advisor, Rare Petro Technologies, Inc. and CEO,
Greenway Energy Transformations, LLC

ACKNOWLEDGMENTS

This book was made possible by many contributors; we cannot name everyone but below are some special contributors:

- Doug Johnson, CEO, Optimal Decision Strategies, Inc., was guest coauthor of Chapter 2, Instrumentation and Measurement and Chapter 3, Data Filtering and Conditioning.
- Satyam Priyadarshy, Chief Data Scientist at Halliburton, was guest coauthor for Chapter 2's section on cyber security.
- Chapter 9 on the future of the digital oil field had a number of industry experts who agreed to be interviewed or provide feedback in a survey. For contributions to Chapter 9, we acknowledge: Raed Abdallah, CEO of Evinsys; Kunal Dutta Roy, VP, and Senthil Arcot, Director, both of Technical ToolBox; Chris Lenzsch, Solutions Manager Big Data and Analytics and IoT, Dell Inc.; Steve Dietz, CEO and Cole Wiser, Director, both of 900LBS of Creative; Anthony McDaniel, CEO of Rare Petro Technologies, Inc.; Pallav Sarma, Chief Scientist, Tachyus; Vasilii Shelkov, CEO of Rock Flow Dynamics; Vasily Demyanov, Professor, Heriot Watt University; and Mohammad Askar, Petroleum Engineering Specialist at Saudi Aramco.
- The authors thank Donna Marcotte for editorial services, and Kattie Washington and Kattie Hammon of Elsevier, for their support throughout the process. Special thanks to Philip Embleton of Saudi Aramco for review and edits of the manuscript.

We acknowledge the many institutions, companies, and individuals who provided permission to use photos and graphics from technical papers, white papers, websites, etc., who are acknowledged within the text.

CHAPTER ONE

Introduction to Digital Oil and Gas Field Systems

Contents

World energy demand is expected grow from about 550 quadrillion (Q) BTU in 2012 to 850 QBTU in 2040 according to 2016 projections from the International Energy Agency (IEA). As can be seen from Fig. 1.1, although renewables will grow by a large percentage, petroleum-based liquids (oil) and natural gas will continue to be the largest contributors to energy utilization by the world's population, representing about 55% of the total. Since any current oil and gas production naturally declines, the continued growth of petroleum fuels will be made possible only by leaps forward in technology in finding, drilling, and producing those resources more efficiently and economically. One of the great stories in oil and gas

Intelligent Digital Oil and Gas Fields
https://doi.org/10.1016/B978-0-12-804642-5.00001-3

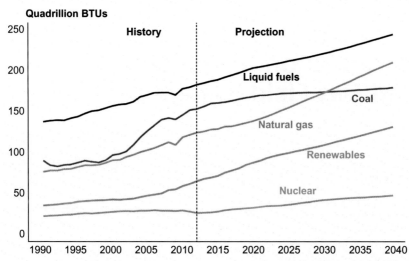

Fig. 1.1 World energy consumption by energy type [International Energy Outlook 2016, Report Number: DOE/EIA-0484(2016). Projection of 46% increase over base 2012].

production is the industry's implementation of new digital technologies that increase production for less unit cost. This "revolution" of the "digital oil field" (DOF) is the subject of this book.

An oil or gas production field is a value chain of a well drilled into a subterranean rock formation from which oil or gas flows to perforations in the well, and then up the well through various chokes and valves to surface pipelines and treatment facilities and ultimately to a sales terminal or tank. Fig. 1.2A is an offshore complex of wellheads and production facilities, and Fig. 1.2B illustrates a complex of pipes from an offshore complex to wellheads on a sea floor, which connect to wells drilled miles under the seafloor to an oil reservoir. Fig. 1.2C illustrates wellheads on an onshore complex which uses pumps to lift the oil. For each of these situations, there are numerous measurements that enable the operations and production personnel to monitor and improve the production (Fig. 1.2D). These measurements include pressures, temperatures, flow rates, power, vibration, and many more to be discussed in this book. DOF is the application of state-of-the-art technology for sensors, data communication, data analytics, collaboration, and decision making, throughout the value chain of completions, wells, pumps, pipes, chokes, compressors, treatment facilities, etc. to increase production at lower cost.

This book leads through the journey of how DOF came to be, the state of DOF today, best practices that can be employed in your own professional work—whether management, operations, or production—and a view of

Fig. 1.2 (A) Offshore oil production platform, (B) subsea well production template, (C) onshore production from artificial lifted wells, and (D) operations center.

what will be possible in the near future. This chapter provides an introduction to DOF and the subsequent topics to be covered.

1.1 WHAT IS A DIGITAL OIL AND GAS FIELD?

When many people first hear the term *digital oil field* they visualize things such as computer displays where you can drag-and-drop digital objects and push buttons to monitor and automate the equipment used in oil and gas field operations. This common vision most definitely describes the *interface* for a DOF system. But behind the crystal screens are many sensors, cables, circuits, electronic switches, logic algorithms, and computers, which are integrated with the specialized instrumentation and equipment—below ground and on the surface—required to run oil and gas fields. This sophisticated collection of technology is further integrated with processes and people into a single system that allows people to interface with the computer technology to optimize the operations of oil and gas fields. It is this collection of technology, processes, and people into an intelligent system that we refer to as the DOF.

DOF systems are commonly known for the promise of delivering the right data, to the right people, at the right time for effective asset decision making that supports a company's objectives in terms of maximizing hydrocarbon recovery and improving operational efficiency. Several common definitions from the literature include the ones listed below:

- "Digital oil field is an umbrella term for technology-centric solutions that allow companies to leverage limited resources. For instance, such technology can help employees to more quickly and accurately analyze the growing volumes of data generated by increasingly sophisticated engineering technologies (Steinhubl et al., 2008)."
- Saputelli et al. (2013) have defined DOF as "the orchestration of disciplines, data, engineering applications, and workflow integration tools supported by digital automation, which may involve field instrumentation, telemetry, automation, data management, integrated production models, workflow automation, visualization, collaboration environments, and predictive analytics."
- "A digital oil field is defined by how a petroleum business deploys its technology, people, and processes to support optimizing hydrocarbon production, improving operational safety, protecting the environment, maximizing, and discovering reserves in addition to maintaining a competitive edge." (istore https://www.istore.com/).

We define a digital oil field implementation as a technology system that integrates high-volume data acquisition and transmission in real time for using data in operations centers, distributed computer systems, and mobile technologies. From these destinations, data are reproduced in virtual models and visualized in a cross-discipline collaborative environment by automated workflows, machine-to-machine communication, intelligent agents, and predictive analytic systems. An integrated DOF approach enables a company to maintain its oil and gas operations at optimal and safe operating conditions and ultimately maximize financial potential with minimum human intervention.

Fig. 1.3 depicts the most important elements of a DOF system.

1.2 DOF KEY TECHNOLOGIES

Many in the industry would argue that DOF began in the early 1980s with the introduction of SCADA (supervisory control and data acquisition) systems in oil and gas production. Those first programs were analog systems with charts that were manually read and interpreted by people. The term DOF is more appropriately associated with introduction of digital sensors,

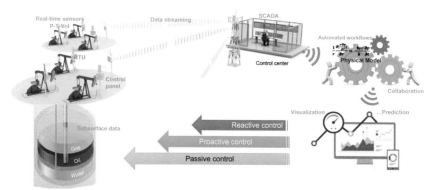

Fig. 1.3 Main description of a DOF system, showing the value chain process from data acquisition, transmission, recollection, data processing, virtual models and workflows, collaboration, visualization and model prediction, and ultimately an action plan with different control modes.

data communication, and computation the mid-1990s, which was followed by an evolution of data and computation in the 2000s. In the mid-2010s, we argue that DOF is in adolescence, with growing "big data" technologies, increasing automation and intelligence being applied, introduction of collaboration decision centers, and emphasis on work processes.

But a golden era of DOF—with sensors on all relevant equipment mobility of applications, intelligence in automation, and automated optimization of production and field management in real time—is on the horizon in the coming years and decades. We can say with confidence that the coming decade will introduce exciting new advancements. Let us take a look at some of the key technologies that were crucial to getting us this far and will be crucial to achieving this golden age of DOF systems.

What we now know as the Internet began through a partnership between the military, universities, and private corporations known as the ARPANET (Isaacson, 2014). Isaacson (2014) provides a history of the full development. The government-sponsored ARAPNET of the early 1960s eventually enabled multiple computer connections using packet switching and distributed network hubs. ARAPNET went from strictly government, to a network of academic institutions in late 1960s, to a commercial enterprise and a standard protocol (IP/TCP) in the mid-1970s. By the late 1980s the Internet began connecting the world and, with standard Internet protocols (IP), enabled DOF systems to connect sensors throughout the oil and gas value chain to centralized, distributed, and mobile computers.

At the start of the new millennium, more than one-third of the total global population used mobile phones transmitting an astronomical record of 180 exabytes in data such as text, pictures/videos (low resolution), and audio. In 2007, smartphones (iOS and Android) were launched, officially ending the analog era, and open many windows into cyberspace and hand-held portable devices (not a desktop computer).

In the last 10 years, the era of *cloud computing*—where data are stored in a repository data center to maintain structure, organize, and process and are accessed through public or private networks—has become endemic. Today, the common expression for the immense data size with exponential expansion is *big data*, a reference to the fact that massive volumes of digital data are not only stored but beyond that the data represent interconnected sources and the data are actually analyzed (by machine learning, neural networks, and process statistics) in real time to enable a multitude of business decisions and transactions. These systems rely on sharing of resources to achieve data coherence and economy of scale.

With a global network and massive amounts of data available, both consumers and businesses want access to all of that with more than just a hand-held device. The *Internet of Things* (IoT) (Fig. 1.4) is the term used to describe connecting a series of devices integrated with electronic, software, and sensors to the Internet to allow sensing and controlling remotely any object such as home alarm, doors, heating and cooling system, cars, etc. In 2015, Cisco announced that more than 99% of total objects in the physical world are still not connected to the Internet (Evans, 2012). They predict that by 2020, 37 billion "smart things" will be connected to the Internet. A Wall Street Journal article (Kessler, 2015) described how ever-smaller technology will have revolutionary new applications such as releasing sensors into blood streams to detect disease, sensors on glasses projecting directly to the eye's retina, 3D printing of equipment, construction sites customized in real time, and having thousands of sensors in oil wells miles below the ground.

Sensors in oil and gas wells, pipelines, processing equipment, and compressors are becoming much less expensive for the hardware, data transmission, communication, and their deployment compared with systems just a few years ago. Data acquisition that used to require expensive instruments, terminals and panels, communication lines, and slow transmission can now be done for a fraction of price, equipment footprint, and data limits. Thus technology advancements were key to enabling the DOF a reality, as you will see in the following chapters.

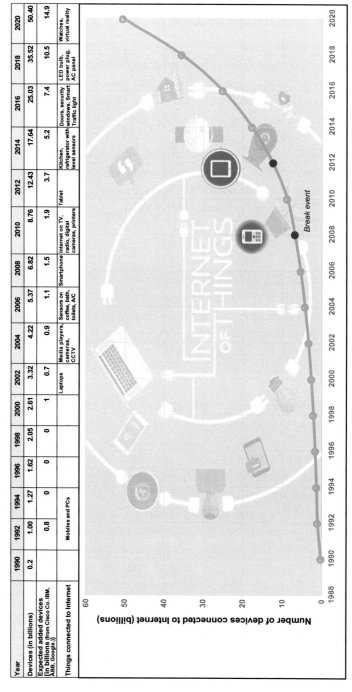

Year	1990	1992	1994	1996	1998	2000	2002	2004	2006	2008	2010	2012	2014	2016	2018	2020
Devices (in billions)	0.2	1.00	1.27	1.62	2.05	2.61	3.32	4.22	5.37	6.82	8.76	12.43	17.64	25.03	35.52	50.40
Expected added devices [in billions (from Cisco Co. IBM, ABB, Google.)]		0.8	0	0	0	1	0.7	0.9	1.1	1.5	1.9	3.7	5.2	7.4	10.5	14.9
Things connected to Internet		Mobiles and PCs					Laptops	Media players, cameras, CCTV	Sensors on coffee, bath, toilets, A/C	Smartphone	Internet on TV, radio, digital cameras, printers	Tablet	Kitchen, refrigerator with level sensors	Doors, security windows, Smart Traffic light	LED bulb, power plug, AC panel	Watches, virtual reality

Fig. 1.4 The Internet of Things (IoT) describes the trend to network all devices for consumers and industry. By 2020 experts expect more than 37 million smart things to be connected to the Internet.

1.3 THE EVOLUTION OF DOF

Beginning in the mid-1990s a series of key projects defines the evolution of DOF systems. This section introduces these projects and their respective highlights with regard to the oil and gas operations that DOF technologies have been applied to. Tables 1.1–1.9 in Section 1.8 summarize each of these projects including the value reported by the operator (which is discussed in this section).

In 1996, Statoil and the Norway Scientific Council (Norwegian Oil Industry Association, 2006) developed a program called integrated operations (IO) to exploit deeper subsea fields under a new information and communication technology program with goals to

- achieve zero environmental accidents,
- minimize human intervention and exposure to high-risk and remote areas, and
- maximize monetary value for the company.

In 1998, Shell and WellDynamics (Ballengooijen, 2007) introduced the concept of *smart wells* by installing a series of remote-controlled downhole control valve (choke) devices and monitoring production in real time. Later, the project was expanded, including production operation from multiple reservoir and operational complex activities, which are referred to as Smart Fields.

In 2000, BP (Reddick et al., 2008) made a significant advancement in DOF for production optimization when the company invested heavily in fiber communications and established advanced collaborative environments, with monitoring centers based onshore that enabled experts to work directly with offshore operations personnel using real-time information in a program called Field of the Future.

From 2006 to 2011, Conoco Phillips initiated its IO (integrated operations) program in the North Sea and Norwegian continental shelf (Digital Energy Journal, 2006). The program focused on operations, an engineering toolkit, and data management. For selected assets the operations staff put all artificial-lift wells on SCADA, constructed an integrated operations center (IOC), and instituted workflows for alarm management, tank management, visualization, and cost optimization. Data management consisted of ensuring a wellhead record and transition from distributed data systems to a single "source of truth" system. The IO program introduced management by exception and tracking of well performance against a plan, which was charted and displayed in the IOC. The IOC was also equipped to control wells and facilities. The IOCs enabled proactive management and addressed issues such as production bottlenecks.

In 2005, Chevron rolled out its I-Field program (Oran et al., 2008). For the San Joaquin Valley Business Unit in California, I-Field projects included collaborative environments (decision support centers, DSC), remote collaboration and visualization, and standardization. The DOF system yielded increases in crew efficiency, better integration of office and field activities, optimization of steam systems, surveillance of well events, and a pattern-exception tool. For its Agbami development in Nigeria (Sankaran et al., 2010; Ibeh et al., 2015), Chevron implemented intelligent well completions, management by exception, improved collaboration, standardization and centralized analytics, and increased focus on safety and environmental risks. The full DOF implementation in Agbami was a critical success factor in the reliability of intelligent wells and minimizing interventions (Ibeh et al., 2015). The Agbami DOF system uses downhole sensors, DSC, and sophisticated data capture and satellite communications. DOF has also been critical for surveillance and flow assurance. In conjunction, a production optimization and reservoir management solution was deployed (Paulo et al., 2011) with well test validation, well rate estimation, and data analytics components.

In 2010, the Kuwait Oil Company (KOC) launched the Kuwait Intelligent Digital Field (KwIDF) program with three pilot projects (Dashti et al., 2012; Ershaghi and Al-Abbassi, 2012), one of the most ambitious in the industry, using state-of-the-art communications, sensor devices, collaboration centers, and automated engineering workflows. The KwIDF vision was to achieve IO for measurement, model, and control of oil field assets, where informed decisions are made effectively and consistently in a collaborative work environment for production and reservoir management. Al-Abbasi et al. (2013) described that DOF is designed to help asset teams meet these challenges, a new generation of petroleum workflow automation integrates real-time data with asset models, helping team members to collaborate so that they can analyze data better and more fully understand the asset problems. They called this program intelligent workflows or *smart flows*.

This approach is cutting edge but also more complex. The complexity is addressed with the use of artificial intelligence technology, such as proxy models and neural networks, coupled with a visualization engine to provide an effective visual data-mining tool. The objective of their workflow automation is to provide integrated solutions to asset opportunities and guide the operations with instructions based on smart analysis and integrated visualization.

Saudi Aramco's first use of an intelligent field program (I-Field), started in 2006 with its Haradh III Increment project, which included multilateral wells equipped with smart completions and real-time data (Al-Hutheli et al., 2012). The program objectives include the following:

- enhancing recoverable HC through in-time intervention and real-time full-field optimization,
- enhancing HSE through remote monitoring and intervention, and
- reducing operation costs by minimizing manual supervision and intervention.

As of 2010 (Abdul-Karim et al., 2010) Saudi Aramco had 19 intelligent fields in operation, with a goal to implement the intelligent field concept in all of its upstream operations by 2017, so it can better understand reservoirs and improve.

1.4 DOF OPERATIONAL LEVELS AND LAYERS

We categorize DOF implementations using the following criteria:
- Are oil wells and related facilities installed with sensors and telemetry?
- What is the level and sophistication of process automation?
- What main type(s) of engineering activity is requested (monitoring, diagnostic, and optimization)?
- What type of working environment is required?

Fig. 1.5 shows how the operational levels of an oil and gas (O&G) business unit impact the major goals of O&G operations.

The pyramid has four layers, with the lowest level of manual processes, increasing to automation, real-time operation center (RTOC), and ultimately DOF. The pyramid also has three vertical axes that represent the key performance indicators (KPIs) as automation increases toward a true DOF implementation; these KPIs are improvements in production uptime, and team and process efficiencies. The efficiency is a value relative to the manual process. Note that other factors are hiding as human intervention and collaboration, where we assume that manual processes are executed completely by human action (i.e., little to no automation) and collaboration is minimal.

In practice, any given O&G operation may operate at several of these levels simultaneously, depending on factors such as the age and size of an operation or asset. For example, older land-based operations may be completely or mostly manual. Newer operations may have more automation, perhaps some SCADA or partially automated with some sensors and other newer or offshore facilities (where risks and costs are inherently

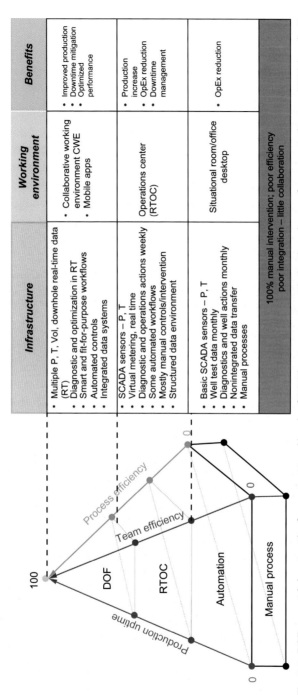

	Infrastructure	Working environment	Benefits
	• Multiple P, T, Vol, downhole real-time data (RT) • Diagnostic and optimization in RT • Smart and fit-for-purpose workflows • Automated controls • Integrated data systems	• Collaborative working environment CWE • Mobile apps	• Improved production • Downtime mitigation • Optimized performance
	• SCADA sensors – P, T • Virtual metering, real time • Diagnostic and operations actions weekly • Some automated workflows • Mostly manual controls/intervention • Structured data environment	Operations center (RTOC)	• Production increase • OpEx reduction • Downtime management
	• Basic SCADA sensors – P, T • Well test data monthly • Diagnostics and well actions monthly • Nonintegrated data transfer • Manual processes	Situational room/office desktop	• OpEx reduction
	100% manual intervention; poor efficiency poor integration – little collaboration		

Fig. 1.5 DOF pyramid showing the four operational levels of O&G business units with the three key performance indicators (KPIs) (production uptime, and team and process efficiencies) that generate value to business. The pyramid is correlated with a table that describes the infrastructure, working environment, and benefits of each operational level.

higher) may be fully automated with many sensors. Below is a description of the four levels in Fig. 1.5.

Manual process level: This is the initial reference point, which assumes a fully manual process. The data are not being acquired in real time. Basic surface sensors [pressure (P), temperature (T), and volume (V)] are used, and readings are recorded and annotated by people when they visit the operation site. Production well tests are conducted periodically, on demand. The data are exchanged between different disciplines by email or in a repository of shared folders. The engineering workflows are performed manually by each discipline (in silos, no integration). Monitoring is performed monthly and diagnostic and optimization is performed randomly (2–3 times a year). Communication is by phone, email, and meetings. Asset team collaboration is low.

Automation level: The real-time data is gathered only from basic surface sensors (P, T), and the information is gathered using wireless technology. Production well tests are taken monthly or on demand. Data are centralized in a SCADA/historian storage center using industry-standard protocols. Some engineering workflows are automated. Monitoring is performed daily, while diagnostic and optimization processes can be performed monthly. Team discussions of production issues are conducted in meetings and situation rooms. Collaboration starts to improve.

RTOC level: Most surface locations in a field have real-time sensors, including flow meters. For wells without flow meters, virtual metering is used for the entire field that can provide production data for individual wells. Data are sent using wireless or WiMAX technologies (to support high-volume data traffic). The data are centralized in the SCADA/historian storage center using the industry standard protocols. Most engineering workflows are automated, with advanced algorithms to provide alarms and alerts. Monitoring is performed in real time, while diagnostic and optimization can be performed weekly or monthly. The operation includes a dedicated real-time operations center, with dedicated staff. Collaboration is significant but not optimal. Communication with field operations staff is via cell phone and by texting.

DOF level: The operations is exactly the same as an RTOC; however, most engineering workflows are intelligent and with predictive capability to generate advice and guidance. Monitoring is performed in real time with exception-based surveillance, while diagnostic and optimization could be performed daily with an advisory system to prevent production downtime. There is a dedicated collaboration working environment (CWE) with dedicated staff and complete workflow mobile communication with field operations staff. Collaboration reaches very high levels with synergy between disciplines. Communication with field operations staff is via closed-circuit TV, video, and chatting.

1.5 MAIN COMPONENTS OF THE DOF

Holland (2012) described three major and essential components for DOF adoption: (1) work processes, (2) technology architecture, and (3) organization (people). These three major components aim in one direction to achieve a company's vision and strategy goals, measures, and incentives. As any large project, DOF is a confluence of people, technology, and process. The main components are then overlaid in what we call the core of DOF (Fig. 1.6), which has five main areas that must be fully synchronized to implement a successful DOF solution: sensing and control, data management, workflow automation, visualization, and collaboration.

1.5.1 Instrumentation, Remote Sensing, and Telemetry of Real-Time Processes

This area focuses on the equipment and technology in the physical oil and gas operations, both on the surface and downhole, required for telemetry, the remote collection and transmission of data required to monitor, optimize, and automate operations. The wellhead includes a series of mechanical or electronic devices (gauges) to measure in real-time pressure, temperature, fluids, and other special data such as chemicals, solids detection, and radiation (Fig. 1.7). Downhole locations are equipped with another family of sensors specially designed to work in high-temperature and high-pressure conditions. Sensors are connected to electrical cables that send analog pulses to a control panel located close to the wellhead.

The control panel consists of many hardware components for the analog-to-digital signal conversion. A key component includes remote terminal units (RTU) and programmable logic controllers (PLC), which perform similar functions. They are connected to sensors with cables, and they send digital data to the transmission hardware using wireless equipment that includes Ethernet, switchboards, WiMAX (microwave signals), and routers all connected to a CPU, which is often powered by a solar panel. The router sends the digital data to SCADA as shown in Fig. 1.8.

Chapter 2 describes of these equipment in detail.

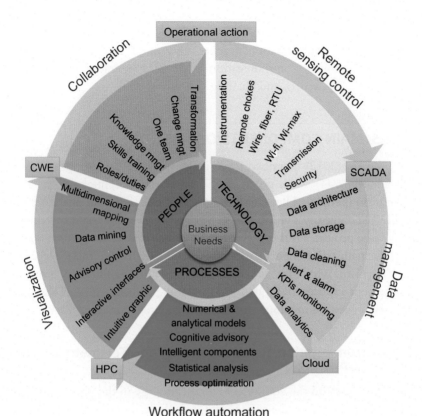

Fig. 1.6 The main components of DOF. The figure shows a business-centric model where the key areas of people, process, and technology are further organized into five critical components of DOF (shown on the outer ring).

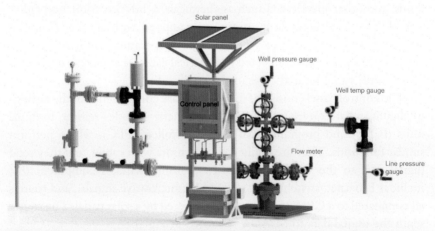

Fig. 1.7 Sensors used to measure pressure, temperature, and flows in real time are connected to a control panel near the wellhead. *(Image courtesy of Rockwell Automation.)*

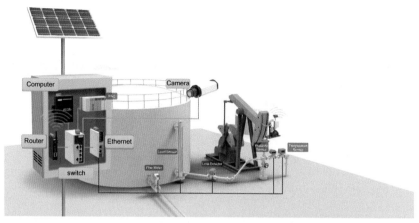

Fig. 1.8 A common configuration for production operations telemetry equipment at a wellsite. *(Image courtesy of Moxa.)*

1.5.2 Data Management and Data Transmission

Located in the SCADA terminal, real-time signals from the field are gathered by cellular modems and sent to a family of servers. The servers use multiplex software to organize and store the data in different structured layers under a series of information technology (IT) industry protocols. The software that does this data collection and aggregation is referred to as a *historian*, which accumulates time data, Boolean events, and alarms in a database, which can be used for many visualization solutions. The data are previously QA/QCed, cleaned, and conditioned using a series of algorithms (data reduction, wavelet filtration, and missing data interpolation) that filters the data from signal abnormalities such as noise, spikes, outliers, and frozen data. The historian commonly feeds a repository or master database, such as the Structured Query Language (SQL) or Oracle. Other types of data—such as mechanical equipment, interventions, tubing scans, gyro scans—are unstructured and stored in well files are well databases.

Chapter 3 discusses these and other processes and key concepts for DOF data management, including

- data architecture, data storage, and cloud servers;
- data QA/QC, conditioning, and cleansing; and
- alert and alarm management.

1.5.3 Workflow Automation

Traditionally, geoscientists and various engineering disciplines (production, reservoir, facilities, etc.) spent considerable time gathering data from disparate sources for input into their mostly manual workflows. Engineers generally use models developed in commercial software applications to reproduce the oil production process. However, even these software models required complex manual workflows that consumed engineers' time, for example: collecting data from different sources (spreadsheet, text, tables, figures, historian, etc.); filtering data from noise; performing repetitive, error-prone tasks to update models (e.g., manual data entry); reconciling the data and calibrating the model; and running different scenarios of the model.

Workflow automation uses high-level programming language routines to connect these manual processes, so that models can be automatically populated and updated. Automation is just part of the DOF requirement for workflow construction. DOF solutions also require that engineering workflows are intelligent enough to capture in real time alarms and alerts to generate prompt actions, update engineering applications, and deliver right-time monitoring, diagnostics, and process optimization that deliver operations guidance at the field level.

Moreover, the workflows should have a predictive character and capability to foresee future operations issues. For these complex tasks, DOF workflows must include sophisticated language program, like artificial intelligence components such as pattern recognition, fuzzy logic, neural networks, proxy models, and optimization supported with advanced multivariate statistical analysis that can generate reliable short- and long-term forecasting. Chapter 4 discusses the concepts of these data analytics.

Chapter 5 discusses the main components of workflow automation, which includes these key concepts:
- Workflow (WF) foundation and philosophy.
- WF types, such as single, integrated, automated, smart.
- Workflow focus including well-centric, task-centric, KPI-centric, and facility-centric.
- Factors that control WFs, such as data- versus model-driven WFs.
- Physical models such as empirical, analytical, and numerical models to serve data reconciliation.
- Virtual models such as virtual metering system when actual metering is not available.

- Predictive models for statistical analysis and intelligent components.
- Cognitive advising such as exception-based management, knowledge capture, and continuous improvement.

1.5.4 User Interfaces and Visualization

Real-time operations centers (RTOC) have traditionally had large screens with dashboards with various graphs, charts, tables, and gauges displayed with real-time data on primarily desktop applications. The newest generation of DOF implementations have displays that go beyond traditional dashboards, and include highly interactive dynamic and map-based, multidimensional (color, size, type elements) displays on Web portals and mobile infrastructure. The displays are fit-for-purpose, focused on individual roles and comprehensive. The user interfaces (UI) display not only the basic data, but also the diagnostics and analytics related to operational actions. The visuals engage and stimulate engineers into collaborative processes. The UIs are designed to guide engineers through a workflow, with automated access and use for all the relevant data (Fig. 1.9). The DOF UIs should be designed with the following features:

- efficient data reload, and data refresh,
- intuitive graphical design, self-explained, auto-flow, no manual requested for simple task,
- GIS map-based feature set,

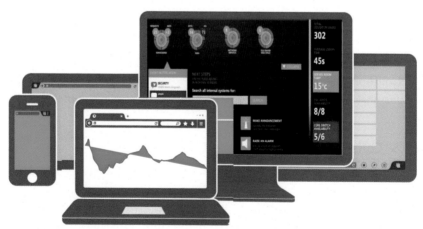

Fig. 1.9 DOF systems have user interfaces that bring together data from multiple sources and display it across multiple devices such as high-end computer monitors, tablets, and smartphones.

- simple and concise with the ability to clear the UI of ambiguous and redundant data,
- interactive to provide advisory control and recommendation for future actions.
- ability for data mining and infographics to show results of clustering, correlation, and data classification. Ability to show results in multi-dimensional maps, plots, and graphics, and
- mobile UI to access data at anytime, anywhere, and from any device (smartphone, tablet, laptops workstation, collaborative work environment).

Traditional operation centers with multiple large monitors were initially designed to monitor a single engineering focus (e.g., drilling or facility operations) and were staffed by appropriate discipline experts. DOF is evolving into new physical spaces called *collaboration work environments* (CWE) or *decision support centers* (DSC), which are designed for collaboration from multiple disciplines using fit-for-purpose workflow visuals that integrate across an operational value chain. These spaces can be categorized (depending on their business operations) as meeting rooms, situation rooms, RTOC, and CWE (which is typically considered the most advanced). The visualization requirements for monitors and screens are to display in real time, multiple sources of raw and processed data.

Chapter 8 discusses the requirements for successful visualization, including the following:

- Hardware such as computers or high-performance computers (clusters), servers, fiber optics, Ethernet, large-screen high-definition (HD) monitors, cameras, projectors, communications, etc.
- Internet with broadband to allow fast emailing, real-time messaging, screen sharing, and videoconferencing.
- Collaborative workspace to allow open visualization with easy-access personal monitors for closer (re)view of a large screen's contents.

1.5.5 Collaboration and People Organization

Traditionally, disciplines involved in the reservoir management value chain, worked in discipline silos, multiple manual data handoffs, use of different systems, and inefficient communication. A successful DOF system requires collaboration among all the disciplines involved in this value chain, including those professionals in remote locations. Technology today allows videoconferences with all team members regardless of their geographic location.

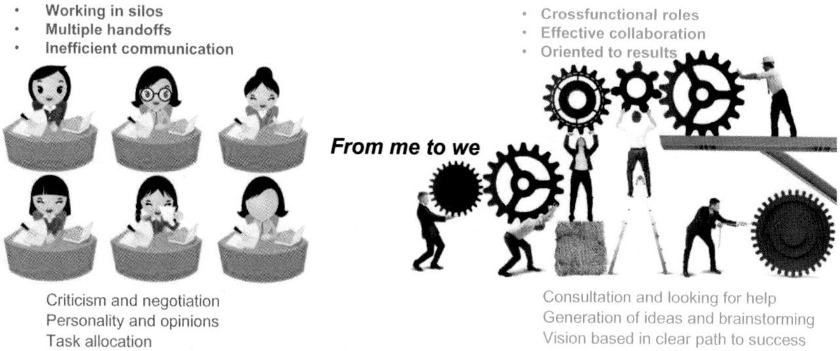

- Working in silos
- Multiple handoffs
- Inefficient communication

- Crossfunctional roles
- Effective collaboration
- Oriented to results

From me to we

Criticism and negotiation
Personality and opinions
Task allocation

Consultation and looking for help
Generation of ideas and brainstorming
Vision based in clear path to success

Fig. 1.10 An example of organizational transformation.

Fig. 1.10 shows how a traditional, siloed organization can be transformed into collaborative teams.

Collaboration implies being connected with the operation center, engaged in an open dialogue, contributing to the discussion, cooperating with others to find solutions, and communicating the results. However, the main organizational challenge is adequately preparing the team for the changes that occur during the transition to and implementation of DOF systems. In this sense, oil and gas organizations must establish training and change management plans, for both newer and more experienced professionals, to acquire the necessary skills for working in a DOF environment. Chapter 8 discusses the tasks and programs required to prepare properly the work teams; these include:

- Identify skill sets and define roles and responsibilities for all professionals and staff, not only in their own disciplines but also their contributions in the collaboration arena.
- Provide mentoring and coaching by experienced professionals.
- Establish a clear and effective chain of command.
- Establish and maintain a high-performance, professional environment, and to specify the appropriate behavior for working in the new environment (e.g., cooperative, respectful of other opinions, etc.).
- Recognize the need to manage stress, particularly for 24/7 real-time emergencies, and to manage different temperaments and across different cultures.
- Establish clear processes to mitigate and manage risks.
- Set clear processes and goals for change management, asset transformation, and continuous improvement.

1.6 THE VALUE OF A DOF IMPLEMENTATION

1.6.1 Industry Challenges

Upstream oil and gas is characterized by intense competition for acreage, capital, and markets; so all hydrocarbon producers must focus on production efficiency. Oil prices are volatile and the costs for extraction typically increase in more challenging environments, such as deep water and arctic locations and unconventional resources (e.g., shale oil and gas). Mature fields also require intense focus on operating efficiency.

Environmental concerns are paramount in many locations such as the arctic regions, deep water, and urbanized areas. Thus, there is potential for remote operations with less labor. Even with volatile oil prices, the world demand is poised to increase at a rate of 0.25% a year, which requires an increase equivalent supply of 200 M STB/d of crude per year. This increased demand requires new development and production activity.

In the quest for a better balance between complying with environment policies and meeting the world oil demand, the oil and gas industry faces strategic challenges in these areas:

- complex operations: complexity within a single asset offshore, multiple wells, multiple possible points of failure, well operations, well interventions, completions, and logging occurring miles below the ground;
- potential for significant Health, Safety, and Environment (HSE) impacts: inherently dangerous operations (complex machinery in remote environments), high risk to people; if an event occurs negative impact to the environment);
- global energy demand versus cost to produce;
- reducing human intervention at operation;
- transforming, sometimes high risk, oil operations into modern technology-driven ones;
- renovating the sluggish and reactive oil operations into faster response to the undesired events, with better prediction of malfunctioning by using an automated process;
- improving cross-discipline communication and collaboration; and
- management by exception to reduce intervention costs, labor, and EH&S impacts.

1.6.2 How DOF Systems Address Challenges and Add Value

The best-in-class DOF projects deliver hard and soft cost-benefit analysis for DOF implementation against key performance metrics. For example, reducing operational costs by a certain percentage, decreasing downtime by a certain percentage, increasing staff efficiencies, managing by exception, optimizing production rates by a certain percentage (artificial lift), and optimizing facility operations. There are tangible and nontangible metrics, which can used to evaluate the DOF investment and performance:

With tangible metrics or key performance indicators, the DOF seeks to maximize:

1. Reduce risks and minimize exposure of operators and professionals assisting in a workplace injury. The ultimate objective is achieving the "zero accident" performance. This is invaluable.
2. Increase production uptime or uplift.
   a. Monitor reservoir decline rate or water-flooding process. While DOF will not increase the well production over its maximum production potential, the asset team can monitor in real time if the oil decline rate diverges from the expected engineering calculation (plan). Moreover, with downhole equipment, DOF can detect on time water breakthrough to control well production.
   b. Detect real-time production and rate deviation from plan. In fields with more than 50 wells and using artificial lift, operators need to go up and above to visit locally all the wells in a day. Probably, one-third total wells are shutting down or producing under expected value. There is not time for operators to realize which wells have troubleshooting. With DOF, the operator can visualize that the troubled wells are in trouble and react with the remedy plan.
   c. Determine opportunities to increase production. Real-time data integrated with reservoir, production, and geological models can reveal information that are hidden by simple monitoring of production data.
3. Increase work process efficiency.
   a. Data can be sent in real time to the operation center. Troubleshooting can be anticipated that is proactive by analyzing the signal behavior.
   b. Data are stored in a structured and organized manner in operations database using a standard industrial process, whereas in the manual process the data are saved on personal backup units.

    c. The automation process boosts the entire manual workflow orders of magnitude compared with baseline performance, reducing human error and miscalculations.
4. Increase staff/team efficiency.
    a. Today between 60% and 80% of engineers are spending their time in data collection and management. Why not engage machines in performing this tedious and error-prone task and dedicate the engineering time for identifying short- and long-term opportunities?
    b. Allow more time spent in troubleshooting, addressing performance issues, and generating in right time rapid diagnostic and present short-term opportunities.
    c. Manage by exception with automated advisories ensuring that the staff addresses issues based on value and key performance criteria and in the most efficient order
    d. Focus on long-term opportunities.
5. Decrease other operational costs
Intangible metrics:
- Better collaboration and cooperation among professional disciplines.
- Proactive addressing of operational issuesless reactive
- Increasing teamwork motivation
- Fast engagement and identification (ownership but with governance) with company goals.

1.6.3 DOF Benchmarks Across the World

This section provides examples of some of the business value generated by the DOF projects discussed earlier in this chapter and are summarized in Tables 1.1–1.9. Fig. 1.11 illustrates that since the late 1990s many companies including IOCs, NOCs, and independent E&Ps have implemented DOF around the world. Tables 1.1–1.9 provide a summary of the characteristics for a survey of DOF projects.

1.6.3.1 Smart Fields

To quantify the added value of its Smart Field program, Shell implemented a rigorous value assessment (Van den Berg et al., 2010) based on conservative estimation and calculation rules. The assessment included 50 Shell assets worldwide, evaluating the impact for 12 months (in 2009). The categories evaluated during the test were smart wells, well optimization, facilities optimization, remote/automated operations, and team efficiency

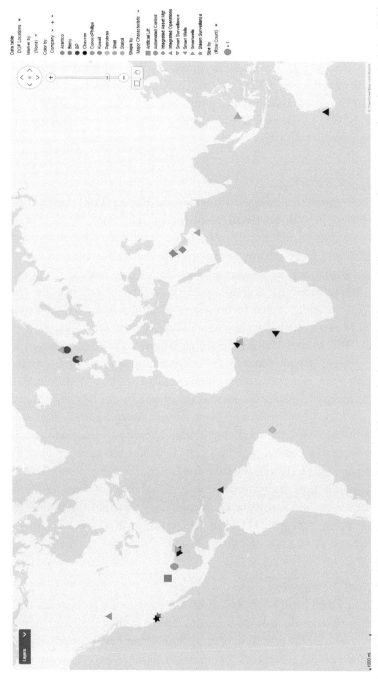

Fig. 1.11 Locations for DOF projects around the world since the late 1990s (the map is illustrative and not meant to be exhaustive). See Tables 1.1–1.9 for DOF details.

improvements. The assessment quantified an overall benefit of $5 per oil barrel when the crude oil price was $30 per STB. The Smart Field program contributed to an additional production of 70,000 STB/d. Reduction in capital expenditure (CAPEX) was estimated at $800 million. After successful implementation of the Smart field program, Shell implemented worldwide an integrated production system model (IPSM) that models and tracks production problems, such as pipeline bottlenecking, well with solid intake, erosional choke, electric submersible pump (ESP) downtime, and separator pressure issues.

1.6.3.2 Field of the Future

BP (2017a,b) stated that in 2007 its Field of the Future (FoF) program delivered between 30 and 50 MBOE/D of gross production. BP has installed 1750 km of submarine fiber optic cable, linking its subsea wells to its worldwide operations. The company has set up a dedicated big data and analytics innovation laboratory in its Center for High-Performance Computing (CHPC) in Houston. BP also has invested significant engineering effort and CAPEX to apply digital technologies to model hydrocarbon flow in its Gulf of Mexico (GoM) facilities to add more production at lower cost per barrel. BP's Model-Based Operational Support (MBOS) program uses real-time surveillance to identify ways of enhancing operating performance; for example, the giant Thunder Horse field in the GoM monitors the buildup of harmful asphaltenes, so that this problem can be mitigated, significantly reducing costly blockages in wells and risers. Sophisticated digital optimization technology is also being used to overcome a common industry problem—unstable flow in pipelines and risers, known as slugging, which can cause platforms to be temporarily shut down. BP's slug controller technology uses real-time measurements and complex algorithms to identify slugs as they form and automatically adjusts choke settings to stabilize flow without stopping production. In this way, costly manual interventions can be avoided and production stabilized.

1.6.3.3 KwIDF Program

After the successful implementation of three major DOF systems, KOC did an 18-month evaluation of the value added by its intelligent systems. Al-Jasmi et al. (2013) and Yunus et al. (2014) reported a significant oil gain of 8% per day, per well by increasing water injection by up to 14 MSTB of water in a total area of 60 wells, in wells subjected to a pattern of waterflooding process. In 12 months, it was monetized for a total of 720 MSTB at $90 per barrel. More important, its intelligent system could predict with acceptable accuracy ESP wells with developing production problems. Jamal et al. (2013), Al-Enezi et al. (2013), and Al-Mutawa et al. (2013) reported that combining proactive and reactive actions—such as increase/decrease of ESP frequency and choke setting, cleaning wells, and shutting in temporary wells—KOC achieved an average of 37% gain in oil production with just 10 actions per well. The total oil was almost 4 MSTB/d with a cumulative production of 756 MSTB in 12 months. Moreover, KOC reported in Al-Jasmi et al. (2013) that the time to analyze one well with production issues was reduced from 7.3 h to only 1.6 h, thereby improving individual and team efficiency. No cost or financial information was reported; however, with these significant oil gains, we believe the operator should have an important reduction in OPEX.

1.6.3.4 Statoil's Integrated Operations

With its IO program, Statoil has realized that the real-time transfer of data over great distances can be used to eliminate the physical distance between installations at sea and support organizations onshore, between professional groups, and internally between the company and its suppliers. A report by Jones (2010) identifies these benefits of IO: improved HSE performance, more efficient drilling operations, better placement of wells, improved production optimization and oil recovery, better reservoir and production control and monitoring of equipment, more efficient maintenance, and increased regularity (production uptime). The benefits have been accounted with 3%–5% increased production, 20%–40% reduction in production losses, and 15%–30% reduction in operating and maintenance costs. In a 2007 report, the Norwegian Oil Industry Association (2007) anticipated that the program might generate an incremental US$41 billion of net present value (NPV).

1.6.3.5 I-Fields

In its I-Field program, Chevron uses developments in sensors, monitoring, and optimization tools that anticipate and plan based on what is happening in real time and continually adjusts to operating conditions. Chevron estimates that its I-Field program has contributed to increases in output and reductions in operating costs by 2%–8% (Meyer, 2010).

1.6.3.6 I-Field Practices

Saudi Aramco has been achieving tangible and intangible benefits from
the implementation of its I-Field practices. The benefits are directly
related to optimizing operating costs, field performance, and safety records
(Abdul-Karim et al., 2010). For example, Al-Malki et al. (2008) report that a
better understanding of reservoir pressure communication during preinjection
stages results in in-time optimum water injection management in new increments, using real-time reservoir management. The I-Field process in the Abu
Hadriyah, Fadhili, and Khursaniyah fields (AFK) helped to reduce planned
preinjection requirements by 14% from the planned volume, over 3 months,
by adjusting the injection rates in the 11 reservoirs in AFK (Al-Khamis et al.,
2009). Additional tangible benefits of I-Field implementation include:

- Significant acceleration of evaluation of extremely complex early inter-
  reservoir communication in the Khurais giant field, between the reser-
  voirs ArabD and Hanifa, now completed in matter of weeks, not years.
- Minimized well interventions, enhanced work efficiency, and timely
  response to production deviations that resulted in an annual fourfold
  reduction in wireline jobs to collect flowing bottom-hole pressure
  (FBHP) and static bottom-hole pressure (SBHP) data (Al-Arnaout and
  Al-Zahrani, 2008).
- Providing quality assurance for crude properties of a blend mixture from
  AFK's 11 reservoirs crude grades at all times (Al-Khamis et al., 2009).
- Enabling a balanced approach in generating data acquisition programs
  and field-strategic surveillance masterplans.

1.6.3.7 COP's Integrated Operations

Conoco Phillips had specific targets, established and tracked KPIs, and
performed evaluations against baselines of its IO program. For its plunger
lift operations, evaluations estimated that annual spending was reduced by
39% in 2012 versus its 2009 baseline (Krushell, 2015). This reduction was
achieved through a centralized management system, smart logic control,
an improved maintenance strategy, and enhanced scrutiny and optimization
of well down procedures. For its intervention processes, the company
reported a 29% reduction in spending for 2012 versus a 2010 baseline,
which resulted from reducing swabbing activity and rod pump downtime
or failure activity and implementing a downhole defect-elimination
process. Other value was captured from improvements or changes in tank
management, operations reporting, trucking logistics, issue tracking, and
data management, and faster communications, which resulted in increased
uptime and production.

1.7 FINANCIAL POTENTIAL OF A DOF IMPLEMENTATION

Companies must evaluate the economic potential of a DOF implementation. The spectrum of investment ranges with the size of a company and its assets: from large international and national companies, to mid-size companies, to even small independents. Costs can range from tens of millions of dollars for the largest assets, to a few million dollars for smaller assets, depending on the sensors, communication networks, IT and database, decision-support requirements, and analytical tools.

The following exercise is a prototype to estimate the value and financial return of a DOF implementation for a 20-year timeframe, for a mid-size, onshore field.

1.7.1 Field Description Example

This exercise assumes an onshore field of 100 wells with a daily production of 20 MSTB of crude oil (30 API), 10 MSTB of water, 20 million SCF of gas, and total recoverable hydrocarbon in situ of 100 million STB. The reservoir is submitted to water injection flooding and the wells produce using ESPs. The oil decline rate is estimated to be 10% annually. This is a new installation, not a retrofit of existing well controls and other equipment.

The operator wants to achieve these goals with a DOF implementation:
- reduce the frequency of well shutdown per month
- increase the oil uplift and decrease oil decline
- delay the water breakthrough and manage water production efficiently

1.7.2 Cost Estimates

These cost estimates are based on price lists from several business cases presented by Cisco, IBM, and Honeywell. Fig. 1.12 shows a breakdown of capital investment costs. For an implementation to reach the goals stated above, an operator would need to upgrade 100 wellheads with controls panels, actuators, choke setting, cables, and pressure and temperature gauges, WiMAX, routers, sensors, and fiber optic cables, with a total cost estimate of $35 million. Additionally, the operator would need to revamp the data center; integrate data streams from the field to the repository center; build servers, CPUs, and cluster with enough data storage for 20 years; and construct a collaborative working environment (CWE), with an estimated total cost of $10 million. To generate new automated workflows and purchase

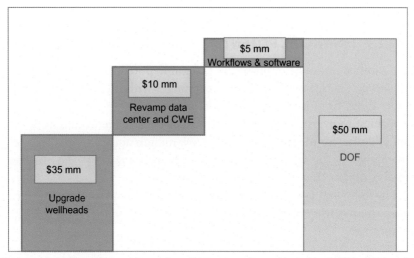

Fig. 1.12 Initial capital investment for a DOF system for an onshore oil field of 100 wells. All currency amounts in US dollars.

software and applications would cost an additional of $5 million. The total $50 million investment, divided by 100 wells is a well cost of $500,000 to build a comprehensive DOF system.

1.7.3 Economic Parameters

The economic parameters for this example include a 20-year timeframe, with fixed prices of oil at $40 per barrel and gas $2.75 per MSCF, a fixed interest rate of 5%, and depreciation of 10% annually. The operating expenses are estimated at $15 per barrel (including salary and wages). The revenue obtained from DOF implementation is estimated by calculating the difference between the annual oil production profiles using a DOF system compared with the production profiles without a DOF system, which is shown in Fig. 1.13.

It is assumed that using a DOF system, ESP downtime has decreased by 70% and also using reservoir analysis, the early water breakthrough has been delayed for more than several months by controlling the water injection. Therefore, it could increase the oil rate by 12%, compared with an oil profile without a DOF implementation.

The initial 50-million dollar CAPEX investment is spent in the first year (red bar). Given this analysis and an oil price of $40.0, the producer can anticipate a total benefit of $88.6 million over 20 years. The NPV is estimated at $64 million and the internal rate of return (IRR) is 22%, when

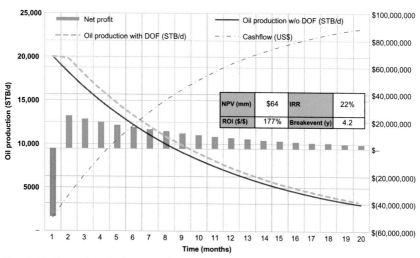

Fig. 1.13 Example of an economic cost estimate and potential ROI of a DOF implementation.

the bank rate is 15% and the return on investment (ROI) is 177%; it is $1.7 gained per $1.0 spent on the project. In Fig. 1.13, the breakeven point is observed at 4.2 years, and the operator can expect substantial economic growth in the following 16 years for a cumulative ROI of around 177%, showing that the DOF implementation could bring significant benefits to the company.

This economic analysis is only an exercise to illustrate the potential benefit of DOF implementation. The costs used here were from older sources. Professionals seeking to do a cost-benefit analysis should seek the current cost estimates and consider the current market and economic conditions and forecasts.

1.8 TABLES SUMMARIZING MAJOR DOF PROJECTS

Tables 1.1–1.9 summarize how companies around the world are implementing major DOF programs, one company per table. The table heading identifies the company, its program name, the first year it reported on its program; the information is from published papers and presentations. Each table has these columns: "DOF Vision and Goals" describes the objective and vision of each company's DOF implementation; "Operations Applied To" includes categories such as drilling, production, operation, production optimization, completions and reservoir engineering;

Table 1.1 DOF Industry Benchmarks for Project: Statoil, Integrated Operation Offshore Real-Time Operations Center (1996) Norway, North Sea, for Fields: Various Fields in the North Sea (Offshore)

DOF Vision or Goals	Operations Applied to	Production Components and Solutions	DOF Best Practices	Automated Workflow Best Practices	Operational Results and Notional Economics
• Zero environmental accident • Minimization of human exposure to risky and remote areas • Maximizing monetary value for the company.	• Production • Drilling • Completion • Reservoir	• Surface field facility • Wellhead and networks • Wellbore and smart wells • Reservoir and geological formation • Completion activity • Monitoring drilling equipment • Real-time production monitoring and visualization • Real time drilling and completion • Production analysis, control, and diagnostic • Production optimization • Planning and scheduling	• Wireless communication • Internet, TCP/IP, routers, RTU, and PLC • Database • Smart wells; ICV/ICD valves • Downhole cables and fiber optic • Distributed temperature and acoustic sensors (DTS, DAS) • High-performance computers • 3D–4D imaging and acoustic signal • Remote operating center • Screen and computers	• Alert and alarm by exception • Automated workflows • Data reconciliation and validation • Data analytic and predictive tools. • Artificial intelligence • Integrated visualization • Integrated surface and subsurface models • Multivariable analysis • Optimization	None is reported

Table 1.2 DOF Industry Benchmarks for Project: Shell E&P, Smart Fields (1998), for Fields: Mark I Nelson Field, North Sea and Worldwide: US, West Africa, Europe/Russia, Middle East, Asia Pacific

DOF Vision or Goals	Operations Applied to	Production Components and Solutions	DOF Best Practices	Automated Workflow Best Practices	Operational Results and Notional Economics
• Monitoring of well production • Hydrocarbon allocation • Forecasting production • Optimizing production system	• Production optimization • Drilling optimization • Brown and green fields	• Wellheads, actuators, ancillary equipment, snake wells, flow lines, and platforms • Clamp-on measurements, wireless transmission • Production modeling using integrated modeling software • Decision/execution done through a collaborative work environment (CWE) • Smart wells using downhole control valves and pressure downhole gauges, and surface control systems. • Integrity monitoring • Rotating Equipment monitoring • Remote Operations • RT surveillance	• Real-time well monitoring and optimization • Real-time well test using fieldware • Well Test • RT process historian • Data acquisition and control • Architecture security • Additional oil-through-gas lift optimization • CWE • Advanced alarming	• Closed-loop reservoir modeling using Petex Software • Integrated visualization • Accelerated system • Implementation methodology for integrated asset management • In-house statistical and data-driven modeling. • Exception-based multivariant alarming • Closed-loop reservoir modeling using Petex Software • Integrated visualization	• Optimized gas lift • Increased oil through production by 15% • Delayed water breakthrough (Brunei) by 2 years • Increased operation reliability by several percent • Reduced manual data by 4h per day • Increased efficiency through automatic data transfer by 2h per day. • Improved operations uptime and reduced production deferment • Smart field value is based on analysis of 50 assets through 2009: $5 billion.

Table 1.3 DOF Industry Benchmarks for Project: BP, Field of the Future (2000), for Locations: North Sea (UK Sector), Angola (Offshore), Kurdistan, Indonesia (Offshore), Trinidad and Tobago, Gulf of Mexico

DOF Vision or Goals	Operations Applied To	Production Components and Solution	DOF Best Practices	Automated Workflow Best Practices	Operational Results and Notional Economic
• Provide the ability to know in real time, equipment status, and optimization opportunities across the production, from the reservoir to the wells, facility, and export. • Optimize production operations by using a sophisticated analytic model, updated with real-time data.	• Drilling and production operations • Facility monitoring • Real-time production surveillance	• High reliable sensors, developing, adopting, and integrating new sensor technologies and transmitting applications. • State-of-the-art equipment for translating insight into foresight through predictive capabilities and advanced visualization. Their main solution is delineated as: measurement, transmission, analysis, interaction and control.	• Use a series of controllers for well downhole and surface. Use fiber optic cable to assure the connectivity between the well downhole and offices. • Use a series of world-class analyzers to generate model and integrate and manage large volumes of information (big data). • Create a family of screen interfaces to facilitate rapid understanding of complex information, which allows better decisions to be made faster. • Use both open/shut-in loop control to execute remote interventions at the right time.	Implemented using the following branch name: • 22 DOF workflows for facility, production operation, and well surveillance. • Production management advisor, a system that allows monitoring well performance in real time, while performing alert and alarm by exception. Uses data mining and predictive analytics. • Operation advisor, a system that allows managing risk proactively, creating alerts for future problems based on expert rules. Monitoring operating envelopes for erosion and corrosion issues. • Well advisor, a system that boosts well efficiency, thereby reducing nonproductive time and improving the life of well integrity.	Digitalization of upstream business has impacted a significant and growing proportion of production and routine support for more than 80% of the 100 top field producers, showing improvement is achieved through Operating efficiency, optimized production, reduced risk, and the ability to make better decisions more quickly.

Table 1.4 DOF Industry Benchmarks for Project: Chevron, i-field (2005), for Locations: Nigeria (Agbami Field and Others), US (Permian Basin), North Sea (Offshore), Australia (offshore), Canada, Angola

DOF Vision or Goals	Operations Applied to	Production Components and Solutions	DOF Best Practices	Automated Workflow Best Practices	Operational Results and Notional Economics
• Maintain plateau production from remote/ offshore areas. • Produce multiple zones through a common wellbore. • Improve reliability and quality assurance • Enhance health, safety, and environment performance (HSE)	• Facilities • Production optimization • RT drilling optimization.	• Jumpers, flow lines, risers, manifolds • Subsea trees and wellheads • Downhole equipment with lower completions • Intelligent well completions (IWC) equipment (densitometers, flowmeters, inflow control valves • Software (Microsoft, multitiered, service-oriented architecture) • Data collection and federation enabled data and application integration, visualization, and logical modeling • Real-time data synchronization system • IWC at wells • Subsea instrumentation	• Integrated and multifunctional team approach to problem solving. • Robust IT architecture and support model. • Proactive system health monitoring and maintenance. • Routine surveillance and operation of wells within constraints. • LPO monitoring • Compression system monitoring.	• 33 DOF workflows facilities, production, reservoir engineering, and field operations. • i-Connect system an integrated diverse data sources and application through open standards and common platform. • i-DOT system an integrated Diatomite Operations—day-to-day planning and execution of thermal diatomite operations. • CiSoft—Sponsor for Center for Interactive Smart Oilfield Technologies (continuous educational program).	• 98% reduction in engineer's NPT/ • Significant time-savings and HSE benefits. • Low-/mid-/high-incremental IWC reserve gain per well of 5/10/18 MMSTBO. • Saved over $500MM-based on proactive real-time optimization workflows. • Optimization of well-rates gained 50,000 BOPD (Agbami). • LPO avoidance saved $10 MM first year.

Table 1.5 DOF Industry Benchmarks for Project: Saudi Aramco, I-Field (2006), for Fields: Qatif, Haradh-III, Khurais Complex

DOF Vision or Goals	Operations Applied to	Production Components and Solutions	DOF Best Practices	Automated Workflow Best Practices	Operational Results and Notional Economics
• Enhance recoverable hydrocarbon through in-time intervention and full-field optimization • Enhance HSE performance through remote monitoring and intervention • Reduce operation costs by reducing manual supervision and intervention	• Production and drilling optimization • Geosteering • Real-time reservoir management • Water and gas injection • Operations and maintenance	• Wellheads • Flow lines • Facilities • Downhole instrumentation • Supply chain solution • Multiphase flow meters • Smart electric submersible pump (ESP) systems • Downhole monitoring systems • Remote-control capabilities • Smart-ready completions (ICV/ICD) • ESP with variable-speed drives (VSD) and pressure sensors • Integrated visualization and engineering systems • Permanent down-hole monitoring systems • RTU with SCADA • Real-time corrosion monitoring	• Real-time actions: fast, strategic and Portfolio interventions • Layered implementation architecture: surveillance, integration, optimization, and innovation	• Data quality control and validation • Process automation • Workflow integration (data and applications) • Collaboration • Knowledge capture • Decision optimization • Interoperability and openness	• Reduced planned downtime and deferral for water injection process and preinjection requirements by 14% from the planned volume • Reduced well interventions four fold, enhanced work efficiency and timely response to production deviations • Optimized operating costs, field performance, and safety records • Accelerated the evaluation of extremely complex, early inter-reservoir communication in Khurais giant field (now completed in weeks, not years)

Table 1.6 DOF Industry Benchmarks for Project: Linn Energy-Berry Petroleum, Reservoir Surveillance Initiative (2012–15), for Fields: (North America) San Joaquin Valley, Mid-Continent Gas; Permian and Uinta Basin

DOF Vision and Goals	Operations Applied to	Production Components and Solution	DOF Best Practices	Automated Workflow Best Practices	Operational Results and Notional Economics
• Maximize oil production • Minimize and mitigate downtime • Reduce well failures • Manage risk of steam surface expressions	• Production • Heavy oil • Steam injection • Gas management	Wellheads, flow lines, and compressors Steam generators, rod pumps. • Surveillance: tilt meters, deviation surveys, mechanical integrity; rod pump dynacards, completion, P, T, multiphase rates. • Data management • Real-time rod pump and plunger lift dynacard surveillance	• Collaboration and control centers per asset • Mobile communications		• Fewer wells down • Increased production and steam efficiency • Reduced compressor bottlenecks. • Faster decisions

Table 1.7 DOF Industry Benchmarks for Project: Conoco, Integrated Operations (2006), Fields: Eagle Ford, Western Canada, Permian

DOF Vision or Goals	Operations Applied to	Production Components and Solutions	DOF Best Practices	Automated Workflow Best Practices	Operational Results and Notional Economic
• Intervention • Production • Cost reduction	Production	• Wells, facilities, and artificial lift. • Plunger lift, rod pumps, facilities, centralized collaboration centers, well automation, and opportunity identification • Alarm management	• Collaboration centers	• Change management and training • Management by exception • Central well control • 24/7 operational surveillance • Pro-active management.	• Reduced well failures • Increased uptime • Improved communication • Reduced costs

Table 1.8 DOF Industry Benchmarks for Project: Kuwait Oil Company, KwIDF (2010) Fields: Lower Burgan, Sabriyah and Gas Field

DOF Vision or Goals	Operations Applied to	Production Components and Solutions	DOF Best Practices	Automated Workflow Best Practices	Operational Results and Notional Economics
• Improve production uptime • Manage the water breakthrough due to water injection • Detect proactively production abnormalities	• Production and reservoir focused workflows	• Wellhead and downhole (DH) wellbore. • Instrumented wellhead with real-time sensors, RTU system, Wi-max, and solar panels • Wellhead with Coriolis meters and multiphase flow meters for well test wells only. • Variable-speed drives on electric submersible pumps (ESP) • Downhole pressure and temperature gauges for ESP wells • Operational database and management • In-house technical software workflows • Collaboration and control center	• Collaboration and decision-support center • Multiphase flow meters • Variable-speed drives on ESPs	• A total of 12 Digital Oilfield Program with more than 10 workflows. • Technical workflows: well performance, well surveillance, and well test management, reporting, gas lift optimization, ESP optimization • Semipredictive workflows to detect water encroachment and production decline	• Improved oil production by 8% per day per well, by increasing the water injection by 30% MSTB of water in a total area of 60 wells • Realized an increase of 37% in overall oil production in just 10 actions/well using proactive actions. • Increased team efficiency by reducing the time to analyze a single well from 7.3 h to only 1.6 h • 12-month monetized value: 720 MSTB of hydrocarbon for an estimated value of t$90 billion

Table 1.9 DOF Industry Benchmarks for Project: Digital Integrated Field Management (GeDIG), Petrobras Fields: Corporate Program for all Petrobras-Operated Fields

DOF Vision or Goals	Operations Applied To	Production Components Solution	DOF Best Practices	Automated Workflow Best Practices	Operational Results and Notional Economics
• Digital integrated field management across all Petrobras operations	• Production losses • Real-time surveillance • Maintenance • Operations centers	• Well instrumentation • Collaboration centers • Automated workflows • Various artificial lift mechanics	• Collaboration • Focus on people, process, and technology model	• Automated technical workflows • Surveillance and operations dashboards • Analytics • Integration of well models • Emphasize change management	• Increased production efficiencies, • Increased recovery factors • Improved operational efficiency

"Production Components and Solutions" identifies the main components and solutions; "DOF Best Practice" refers to advanced hardware and software tools used; "Automated Workflow Best Practices" refers to those software features that allow the automation and the implementation of smart components; "Operational Results and Notional Economic" refers to the available public information with respect to value added by the DOF implementation.

REFERENCES

Abdul-Karim, A., AL-Dhubaib, T., Elrafie, E., Alamoudi, M.O., 2010. Overview of Saudi Aramco's Intelligent Field Program. Society of Petroleum Engineers. SPE-129706-MS https://doi.org/10.2118/129706-MS.

Al-Abbasi, A., Al-Jasmi, A., Nasr, H., Carvajal, G., Vanish, D., Wang, F., Cullick, A.S., Md Adnan, F., Urrutia, K., Betancourt, D., Villamizar, M., 2013. Enabling numerical simulation and real-time production data to monitor water-flooding indicators. In: Paper SPE 163811 presented at the SPE Digital Energy Conference, The Woodland, TX, March 5–7 https://doi.org/10.2118/163811-MS.

Al-Arnaout, I.H., Al-Zahrani, R., 2008. Production Engineering Experience with the First I-Field Implemented in Saudi Aramco at Haradh-III, Transforming Vision to Reality. Society of Petroleum Engineers. SPE-112216-MS https://doi.org/10.2118/112216-MS.

Al-Enezi, B.A., Al-Mufarej, M., Anthony, E.R., Moricca, G., Kain, J., Saputelli, L.A., 2013. Value Generated Through Automated Workflows Using Digital Oilfield Concepts: Case Study. Society of Petroleum Engineers. SPE-167327-MS https://doi.org/10.2118/167327-MS.

Al-Hutheli, A.H., Al-Ajmi, F.A., Al-Shamrani, S.S., Abitrabi, A.N., et al., 2012. Maximizing the Value of Intelligent Field: Experience and Prospective. Society of Petroleum Engineers. SPE-150116-MS https://doi.org/10.2118/150116-MS.

Al-Jasmi, A., Goel, H.K., Al-Abbasi, A., Nasr, H., Velasquez, G., Carvajal, G., Cullick, A., Rodriguez, J., 2013. Maximizing the value of real-time operations for diagnostic and optimization at the right time.Paper SPE 163696 presented at the SPE Digital Energy Conference, The Woodland, Texas, 05–07 March.

Al-Khamis, M., Zorbalas, K., Al-Matouq, H., Almahamed, S., 2009. Revitalization of old asset oil fields into I-fields. In: SPE Saudi Arabia Section Technical Symposium, Al-Khobar, Saudi Arabia, May 9–11 SPE-126067-MS https://doi.org/10.2118/126067-MS.

Al-Malki, S., Buraikan, M.M., Abdulmohsin, R.A., Ahyed, R., Housam Al-Hamzani, H., 2008. I-Field Capabilities Enable Optimizing Water Injection Strategies in Saudi Arabian Newly Developed Oil Fields. Society of Petroleum Engineers. SPE-120835-MS https://doi.org/10.2118/120835-MS.

Al-Mutawa, S.A., Saleem, E., Anthony, E., Moricca, G., Kain, J., Saputelli, L., 2013. Digital Oilfield Technologies Enhance Production in ESP Wells. Society of Petroleum Engineers. SPE-167352-MS https://doi.org/10.2118/167352-MS.

Ballengooijen, v.J., 2007. Smart Field Programme Manager. The Shell Fields Technology Journey. The Hague, Netherland.

BP, 2017a. Digital Technology. http://www.bp.com/en/global/corporate/technology/technology-now/digital-technology.html.

BP, 2017b. Field of the Future. BP.http://www.bp.com/en/global/corporate/technology/technology-now/digital-technology.html.

Dashti, Q., Al-Jasmi, A.K., AlQaoud, B., Ali, Z., Bonilla, J.C.G., 2012. Digital Oilfield Implementation in High Pressure and High Temperature Sour Environments: KOC Challenges and Guidelines. Society of Petroleum Engineers. SPE-149758-MS https://doi.org/10.2118/149758-MS.

Digital Energy Journal, 2006. Integrated operations at ConocoPhillips. Available from: http://www.digitalenergyjournal.com/n/Integrated_operations_at_ConocoPhillips/081eabce.aspx.

Ershaghi, I., Al-Abbassi, A., 2012. A Perspective for a National Oil Company to Transition from Traditional Organizational Management to a Digital Culture. Society of Petroleum Engineers. Paper SPE 150219-MS https://doi.org/10.2118/150219-MS.

Evans, D., 2012. How the Internet of Everything Will Change the World…for the Better #IoE [Infographic]. http://blogs.cisco.com/ioe/how-the-internet-of-everything-will-change-the-worldfor-the-better-infographic.

Holland, D., 2012. Exploiting the Digital Oilfield: 15 Requirements for Business Value, corporate edition Xlibris Corporation LLC, Bloomington, IN.

Ibeh, C., Awotiku, O., Ledegerber, A., Ugah, I., Awonuga, K., 2015. The Agbami Digital Oilfield Solution and Reliability Assessment of Intelligent Well Completions. Offshore Technology Conference. OTC-25690-MS https://doi.org/10.4043/25690-MS.

Isaacson, W., 2014. The Innovators: How a Group of Hackers, Geniuses, and Geeks Created the Digital Revolution. Simon and Schuster, Camp Hill, PA.

Jamal, M.A.-R., Al-Mufarej, M., Al-mutawa, M., Anthony, E., Chetri, H., Singh, S., et al., 2013. Effective Well Management in Sabriyah Intelligent Digital Oilfield. Society of Petroleum Engineers. SPE-167273-MS https://doi.org/10.2118/167273-MS.

Jones, S., 2010. The economic value of integrated operations. In: ABB article of Oil and Gas Petrochemical. Available from: https://library.e.abb.com/public/7cc9654e734fde2dc1257b0900361b07/Integrated_operationsUS.pdf.

Kessler, A., 2015. Shrinking tech means room at the top. https://www.wsj.com/articles/shrinking-tech-means-room-at-the-top-1451345558 Wall Street J., December 28. Available from:.

Krushell, G., 2015. Integrated operations at ConocoPhillips.Presented at the Data Driven Production Optimization Conference, Houston, TX, 16–17 June 2015.

Meyer, M., 2010. Today-value in any part of the cycle, plenary address 1.Presented at the SPE Intelligent Energy Conference and Exhibition, Utrecht, Netherlands, 23–25 March 2010.

Norwegian Oil Industry Association (Oljeindustriens Landsforening (OLF)), 2006. Potential value of Integrated Operations on the Norwegian Shelf. http://www02.abb.com/global/seitp/seitp161.nsf/0/19ff7687080e051dc125718b004b1caa/$file/060521+Potential+value+of+Integrated+Operations+on+the+Norwegian+Shelf%5B1%5D.pdf.

Norwegian Oil Industry Association and POSC Caesar Association, 2007. Integrated Operations and the Oil and Gas Ontology. (accessed 06.05.09).

Oran, K., Brink, J., Ouimette, J., 2008. Implementation Results for Chevron's *i-field* in San Joaquin Valley, California. Society of Petroleum Engineers. SPE-112260-MS https://doi.org/10.2118/112260-MS.

Paulo, J., Taylor, D.A., Isichei, O., King, M., Singh, G., 2011. Transforming Operations with Real Time Production Optimization and Reservoir Management: Case History Offshore Angola. Society of Petroleum Engineers. SPE-143730-MS https://doi.org/10.2118/143730-MS.

Reddick, C., Castro, A., Pannett, I., Perry, J., Dickens, J., Carl Sisk, C., et al., 2008. BP's Field of the Future Program: Delivering Success. Society of Petroleum Engineers. SPE-112194-MS https://doi.org/10.2118/112194-MS.

Sankaran, S., Olise, M., Meinert, D., Awasthi, A., 2010. Realizing Value from Implementing I-Field in a Deepwater Greenfield in Offshore Nigeria Development. Society of Petroleum Engineers. SPE-12769-MS https://doi.org/10.2118/127691-MS.

Saputelli, L., Bravo, C., Nikolaou, M., Lopez, C., Cramer, R., Mochizuki, S., et al., 2013. Best Practices and Lessons Learned after 10 Years of Digital Oilfield (DOF) Implementations. Society of Petroleum Engineers. SPE-16726-MS https://doi.org/10.2118/167269-MS.

Steinhubl, A., Klimchuk, G., Click, C., Morawski, P., 2008. Unleashing Productivity: The Digital Oil Field Advantage. https://www.strategyand.pwc.com/media/file/UnleashingProductivity.pdf.

Van den Berg, F., Perrons, R.K., Moore, I., Schut, G., 2010. Business Value From Intelligent Fields. Society of Petroleum Engineers. SPE-128245-MS https://doi.org/10.2118/128245-MS.

Yunus, K., Chetri, H., Saputelli, L., 2014. Waterflooding Optimization and its impact using Intelligent Digital Oilfield (iDOF) Smart Workflow Processes: A Pilot Study in Sabriyah Mauddud, North Kuwait. International Petroleum Technology Conference. IPTC-17315-MS https://doi.org/10.2523/IPTC-17315-MS.

FURTHER READING

IBM Institute for Business Value, 2010. Oil and Gas 2030: Meeting the Growing Demands for Energy in the Coming Decades. p. 12.https://www-935.ibm.com/services/multimedia/GBE03376USEN.pdf.

iStore, 2015. Digital Oilfield Solution. Istore Co..https://www.istore.com/Overview.html

US Department of Energy, 2010. Predictive Maintenance. http://www1.eere.energy.gov/femp/program/om_predictive.html (accessed 01/2012).

US Energy Information Administration, 2016. EIA projects 48% increase in world energy consumption by 2040. In: Today in Energy.http://www.eia.gov/todayinenergy/detail.php?id=26212.

CHAPTER TWO

Instrumentation and Measurement

Contents

One primary reason for the recent proliferation of DOF systems is the proliferation of DOF infrastructure. As instrumentation, well control systems and supervisory control and data acquisition (SCADA) have advanced and become less expensive, and therefore they are applied more broadly throughout the oil field. As they become more broadly applied, then more data and control is available to use in surveillance, automation, and optimization activities, which are the hallmark of DOF systems.

Chapter 1, Section 1.5.1 presents the elements of the architecture of instrumentation, remote sensing, and telemetry of real-time processes

(Fig. 1.7), and Section 1.5.2 presents data management and data transmission (Fig. 1.8) for SCADA systems. This chapter presents the details of the prevailing trends for well instrumentation, wellhead control, and SCADA systems. It first presents some general trends in each category and is further organized into how these systems are being applied in different asset types. The chapter investigates mature fields, deepwater platforms and floating production storage and offloading (FPSO) systems, and unconventional assets. Although DOF systems are being applied across all oil field activities—from geology and reservoir to drilling, to completions, and finally production—this chapter focuses on production.

Although there are very interesting advances in the areas of database and enterprise IT—for example, data lakes, in-memory data architectures, and open-system analytics to name a few—those technologies are discussed in Chapter 4. In addition, the unique hardware for safety shutdown systems is not presented here. Instead, the hardware and software from the instruments through data historians are discussed. The asset control network discussed here includes field instrumentation, field control devices, telemetry, SCADA systems, and data historians.

Engineers and operators engaged in companies implementing DOF systems have become very familiar with the growth of these systems in oil field operations and are conversant with data-driven production tools. With the concomitant growth of Big Data (Chapter 4) and mobile systems (Chapter 8), field operators and engineers can connect to the field sensor and control systems from multiple systems, for example, networked computers, tablets, phones, etc. from any location to make near real-time decisions.

2.1 INSTRUMENTATIONS FOR MEASUREMENT: GAUGES AND FLOWMETERS

The most important aspect in any DOF systems and automated workflows is to be able to measure well performance in real time. Pressure and temperature gauges and flowmeter and pump measurements are crucial for DOF workflows.

2.1.1 Surfaces Gauges

The most important real-time measurements are surface pressure and temperature. Fig. 2.1 shows a complete set of surface gauges, sensors, and devices useful for DOF operations.

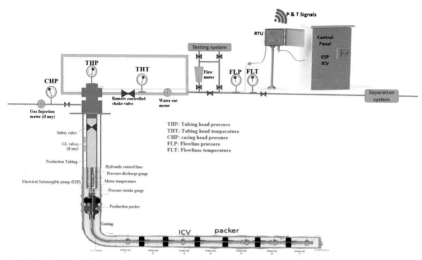

Fig. 2.1 The whole set of gauges, sensors, and meters required for real-time DOF. The figure shows surface and downhole gauges with surface remote-controlled valve and ICVs. An artificial list system is also added; note that the wells can be equipped with GL or ESP but rarely both.

The following are the essential data required for surface monitoring:
- Tubing head pressure (THP) and tubing head temperature (THT) gauges, if wells flow through tubing.
- Casing head pressure (CHP) and THT gauges, if wells flow through casing or annular area, including gas lift in shallow well completions.
- Flowline pressure (FLP) and flowline temperature (FLT).

2.1.2 Downhole Gauges

In dry and wet gas wells, downhole gauges are not required because a two-phase flow correlation (such as that of Gray, 1978; Beggs and Brill, 1973) can be used to estimate the bottom-hole flowing pressure with acceptable accuracy. However, when a multiphase flow occurs and produces gas, vapor, oil, gas-in-solution, and water, then downhole pressure and temperature gauges can generate tremendous value to DOF workflows to measure well and reservoir performance. Moreover, during well shut-in periods, the downhole gauges can capture the essential data to perform a pressure transient analysis (PTA) to estimate the static reservoir pressure (p^*), reservoir conductivity ($k.h$), and skin factor (S).

To be able to measure performance and optimize lift, then wells with artificial lifts must have all necessary gauges installed. In wells using electrical submergible pumps (ESP), generally the ESP is equipped with a downhole intake pressure gauge (inlet before the ESP's motor) and a discharge pressure (outlet after the ESP's motor) plus the motor temperature. Gas lift (GL) wells are equipped with a downhole gas valve, which use differential pressure to estimate the gas injected and the flowing pressure at the tubing. In natural flow wells, we recommend setting up gauges at the end of tubing with a packer; the information is transmitted using electrical cables. Fig. 2.1 shows an example of downhole gauges, sensors, and devices useful for DOF operations. Internal control valves (ICV) are included in this figure; ICV devices are explained in Chapter 7.

2.1.3 Surface Flowmeters

One extremely important trend used at all asset types is the increase in individual well measurement—especially in flowmeter technologies. Any instruments that provide real-time accurate measurements of well production liberate many DOF workflows. There are three ways to develop well flow rate: direct measurement, direct calculation, or virtual measurement using analytic or empirical models. Now operations workflows do not have to depend first on an allocation workflow, which has different objectives and requirements than most operations activities. Comprehensive work on multiphase flow is described by Falcone et al. (2009).

The most common means of measurement is to install a separator followed by individual component flowmeters. Flowmeters have improved beyond the typical orifice plate meters and their prices have fallen. It is very common to find Coriolis meters used for oil or water flows. Orifice plates work well for gas flows. Turbine or Venturi types are also commonly deployed. A major benefit of the Coriolis type, which uses principles of mechanics of fluid flow in a vibrating tube, is that it measures fluid density along with rate so that the operator can tell if gas or even water is flowing along with the oil measured.

However, flow rate instruments have two disadvantages. First, their use requires a dedicated process to ensure that the meter is calibrated and fit for use as the well production declines. For example, Coriolis meters do not perform well at low turndowns and at high gas volume fraction (GVF), and orifice plates need to be changed to fit the flow rate range throughout a well's life cycle. The second disadvantage is that to use separate meters on each fluid stream, one or more separators must be installed. Separators are expensive to purchase and maintain. However many old or even low rate wells are now using individual fluid meters.

Another alternative to separators and individual fluid meters is a three-phase meter. Technologies for these meters are rapidly progressing and include nuclear and sonic-type meters (nonintrusive devices). As with Coriolis meters, pricing has become reasonable for many applications.

2.1.3.1 Types of Fluid properties Measured Over Time and Why

Falcone et al. (2009) described five essential parameters measured (separately) by flowmeters:

- *Meters that measure fluid flow velocity (v).* This category includes classical intrusive meters that have spinners or turbines. A flow passing through the vane of the device moves the spinner; the rate of spin per time is measured to calculate the volumetric flow. Turbine, spinner, or vortex meters are the most common devices that measure the total flow velocity. Most fluids have multiple components (liquid, gas, etc.) The main way to improve precision and accuracy of these devices is to homogenize the fluids and measure total fluid velocity. Under this regime, a turbine meter can measure the total fluid velocity of the mixture but it cannot measure the fluid density.

- *Meters that measure density (ρ).* Nonintrusive devices based on ultrasonic, electrical impedance, or particular atomic constituents can measure the mean density of a fluid on a section of a pipe. This category includes γ-ray adsorption or neutron interrogation. These devices measure the void fraction of the gas and the liquid, but it cannot measure fluid velocity.

- *Meters that measure mass flow ($\rho.v$).* These devices use a flexible tube that measures the amount of mass passing through a u-shaped pipe, causing vibrations; the vibration is converted into total mass flow. This category includes Coriolis meters and true mass flowmeters (TMFM).

- *Meters that measure momentum ($\rho.v^2$).* The product of mass flux (density * velocity) and fluid flow velocity is defined as fluid momentum, or the force of a fluid in motion. This category includes traditional intrusive pressure-drop devices, such as orifices, Venturi, nozzles, etc. The pressure drop is measured directly between the inlet and outlet pressure points.

- *Meters that measure chemical or atomic elements (H_2, C_2, O_2).* This category includes electrical, mechanical, acoustical, hydraulic, or atomic devices that, depending on their physical principles, measure the concentration and velocity of atomic elements such as hydrogen, carbon, and oxygen. For example, infrared spectroscopy water cut device measures the volumetric proportion of oil in a mixture of oil and water by running a beam of infrared light that is absorbed by oil fluid.

2.1.3.2 Flowmeters: Principles of Measurement

The basic types of meters are based on different measurement principles such
as differential pressure, velocity flow, positive displacement, mass flow, and
elemental analysis (ultrasonic, electromagnetic, thermals, radioactive, etc.).

Differential pressure meter. The most traditional meter used by the oil and
gas (O&G) industry, this meter measures the pressure drop over an orifice
inserted into the flow current. Engineers use the Bernoulli equation to estimate the pressure drop, which is a function of the square flow speed multiplied by fluid density:

$$\Delta p = \frac{\rho^* v^2}{2} \tag{2.1}$$

where Δp is the pressure difference measured at the orifice between two
points in psi; ρ is the fluid density in lbs/ft^3, and v is the flow velocity in
ft/s.

The typical devices using this principle are orifice plates, flow nozzles,
Venturi tubes, and rotameters. Fig. 2.2A shows a typical example of a Venturi meter and Fig. 2.2B gives an example of an orifice meter.

Velocity flowmeter: Measures the flow velocity by counting the number of
spins per unit of time (e.g., rotation, revolutions, or spinning per second) and
is multiplied by the cross-sectional area of the pipe:

$$v = \frac{Q}{A} = \#\text{spins}/\left(\pi r^2\right) \tag{2.2}$$

where Q is the flow rate in ft^3/s; A is the cross-sectional area of a pipe in ft^2;
and v is the flow velocity in ft/s. The typical devices using this principle are
turbine meters, vortex flowmeters, and spinner meters. Fig. 2.3 shows an
example of a typical turbine meter showing spin and sensors.

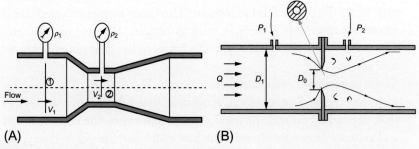

(A)   (B)

Fig. 2.2 (A) A Venturi meter showing inlet pressure at point 1 and outlet pressure at
point 2. (B) An orifice meter showing differential pressure after orifice reduction D_0.

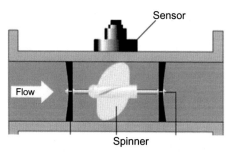

Fig. 2.3 An example of a turbine meter showing the spinner rotating in the direction of the flow.

Positive displacement meter: A series of synchronized gears or rotors is moved by the flow current, which displaces a known volume of fluid over time. The rotations of the rotors are proportional to the volume of the fluid being displaced over a time period. This category includes reciprocal pistons, gear meters, and rotary vane meters. This meter is widely used to measure water in houses. Fig. 2.4 shows rotors synchronized to each other in a piston.

Mass flow meter: Measures directly the mass (molecular weight) passing through a u-shaped tube. The flow mass is a physical property measured independently of its pressure, density, viscosity, and temperature. Commonly, the O&G industry uses a Coriolis mass flow device, which uses the Coriolis principles to measure the quantity of mass moving through the u-shaped tubes that generate a vibration with angular harmonic oscillation; the degree of oscillation is a direct measure of the mass flow. This category also includes thermal flowmeters, which measure the thermal conductivity of the fluid, which is directly proportional to the mass flow. Fig. 2.5 shows an example of a Coriolis meter.

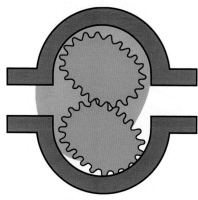

Fig. 2.4 An example of a positive displacement meter.

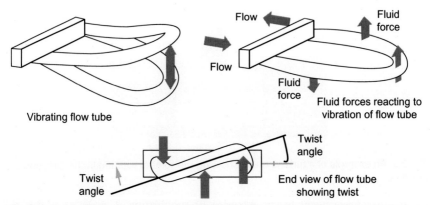

Fig. 2.5 An example of a Coriolis meter. Once the flows pass through the tube, it starts to vibrate and twist. The twist angle is proportional to mass flows.

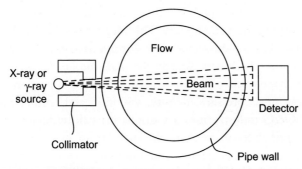

Fig. 2.6 X-ray or γ-ray attenuation or neutron interrogation devices showing beams received by a detector.

Elemental analysis meter. According to Falcone et al. (2009), this meter measures the concentration and velocity of individual atomic elements, such as oxygen, hydrogen, and carbon. The main element types are acoustic, electromagnetic, γ- and X-ray attenuation, neutron interrogations, microwave attenuation, and infrared spectroscopy. Fig. 2.6 shows a nonintrusive elemental meter using X-ray, γ-ray, or neutron source devices to detect the amount of oxygen and hydrogen atoms in the flow.

2.1.3.3 Criteria for Choosing a Flowmeter
- Precision and accuracy of the measurement.
- A meter's operating envelope.
- What are the main fluids produced from the reservoir? Water, gas free, gas in solution, oil vaporized, oil free of gas, heavy oil, water with gas bubble, etc.

- Calibration and maintenance: How often must the meter be calibrated?
- Gas void fraction: How much represent the fraction of gas or oil volume in the pipe through time?
- Cost versus values.
- Government acceptance and health, safety and environmental issues.
- A meter's position before or after a separation system.

2.1.3.4 Key Factors to Consider in Flowmeter Selection

Turndown ratio (TR). TR is the ratio of the maximum observed peak rate divided by the minimum observed rate in a period of time, $TR = Q_{max}/Q_{min}$. It represents the rangeability of the instrument: the higher the number, the better range of sampling. If you have a well producing with slugging patterns, showing intermittent high/low rate in $<1h$, then it requires a meter with high TR, above 10:1.

Flowmeter location. Onshore, traditional meters (mass flow, flow rate, and differential pressure meters) are used after the separation system (low pressure). In this situation, it is very important to evaluate the flow regime and pattern. Offshore, multiphase flowmeters (MPFM) might be used before the separation systems.

Single-phase versus and multiphase flows. Gas is compressible and the density changes significantly with changes in pressure, even when the change is low. Liquids are considered incompressible and, in general, densities (oil and water) do not change with pressure. If the pressure of a wet gas system increases, the density of the gas will increase but liquid density remains almost the same. Compositional fluids such as volatile O&G condensate, change fluid composition with pressure. To distinguish the fraction of phases on the flow rate (oil, water, and gas) requires flowmeter technologies to quantify the volume fraction of each phase. The engineers need to know if the fluids in a particular zone are flowing in single or multiple phases, because it will change the type of meter needed.

Flow regimes and patterns. Flowmeters generally are placed between the wellhead and the separation system or immediately after the separation system. At this stage, fluids are governed by gravitational or kinetic forces. The flow pattern depends on the volume of each phase, vapor and liquid properties, and the pressure and velocity of each phase. Therefore, the phases can be distributed along the horizontal pipe in many sections and in many ways, which is shown in Fig. 2.7 with a description below the figure.

- *Stratified or wavily stratified flow* takes place at low pressure, gas, and liquid flow at different velocities; the surface between the liquid and the gas are

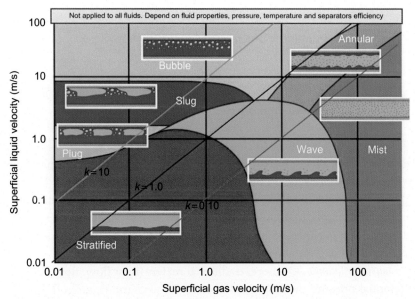

Fig. 2.7 A cross-plot in *Y*-axis with superficial liquid velocity versus superficial gas veloc-ity. The figure shows the different patterns generated in a horizontal pipe. Additionally, relative slip velocity is shown with a straight line between $k = 10$, 1.0, and 0.1. This chart cannot be applied to all fluids; there is a specific chart that depends on fluid properties, pressure, temperature, and separator efficiency. *(Image courtesy of the Norwegian Soci-ety for Oil and Gas Measurement and The Norwegian Society of Chartered Technical and Scientific Professionals, 2005. Handbook of Multiphase Flow Metering, second ed. Oslo, Norway.)*

separated clearly. Several parameters are used to distinguish between single- and multiphase flows. After flow rate, one of the most imprecise properties to be measured is the fraction of gas, oil, and water in a hor-izontal pipe. Graham (2014) described that the fraction of gas occupied in a pipe is called the gas void factor (e_g), which has been one of the main factors to consider for flowmeter design. In fluid dynamics, production engineers use three parameters to distinguish between single- and multiphase flows, such as the gas void fraction, GVF, and multiphase parameters, as described below:

− The first calculation is gas void fraction (e_g), which is calculated using the area of the pipe occupied by gas divided by the total cross-sectional area of the pipe (Fig. 2.8). The e_g method assumes that the fluids flow at a low velocity (laminar) and that gas and liquid travel at different velocities. When e_g is higher than 0.70, it is con-sidered gas single phase (dry gas or wet gas), a value of between 0.4

Void fraction ➤ Volume fraction

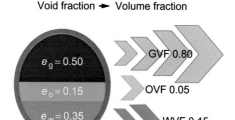

Fig. 2.8 A cross-section of the pipe showing a stratified flow pattern with calculated gas, oil, and water void fractions with the corresponding gas, oil, and water volume fractions.

and 0.7 is defined as multiphase flow (gas condensate, volatile oil, water cut over 30%). Below 0.3 could be considered liquid single phase (black oil with low gas/oil ratio (GOR), heavy oil, high water cut >70%).

— Another important factor is gas volume fraction (GVF$_{vol}$), which is estimated as the ratio of the total gas flow to the total rates of water, oil, and gas. The e_g and GVF$_{vol}$ are unequal and different levels of thresholds. When GVF$_{vol}$ is higher than 0.90, it is considered gas single phase (dry gas or wet gas), between 0.5 and 0.89 is defined as multiphase flow (gas condensate, volatile oil, water cut over 30%). Below 0.5 could be considered as liquid single phase; however, GVF$_{vol}$ and GVF$_{area}$ are related through the following expression 2.5:

If gas and liquid travel at different velocities:

$$Vr = V_g - V_l \,(\text{slip velocity}) \tag{2.3}$$

So slip ratio can be calculated as:

$$k = V_g / V_l \tag{2.4}$$

Assuming that the gas velocity is five times the liquid velocity, then $k = 5.0$ and the fraction of gas occupied in the pipe is $e_g = 0.8$.

The GVF$_{vol}$ is

$$\text{GVF}_{vol} = \frac{e * k}{1 - e + e * k} \tag{2.5}$$

GVF$_{vol} = 95\%$ and the fluid is gas single phase.

Fig. 2.8 shows an example of a pipe flowing three phases with a void gas fraction e_g of 0.50, oil 0.15, and water 0.35. The corresponding gas, oil, and water volume factor is also shown.

- The Lockhart-Martinelli parameter (χ) can be calculated from the mass flow rate or volumetric flow rate of liquid and gas and the density of the fluids, which is a dimensionless number. The parameter helps to define the wetness or liquid loading of the gas using the following range: A wet gas flow has a value of between 0.0 and 0.3, and values above 0.3 are defined usually as multiphase flows.

$$x = \frac{Q_{liq}}{Q_{gas}} \sqrt{\frac{\rho_{gas}}{\rho_{liq}}} \qquad (2.6)$$

In summary:

Phase	Gas void factor (e_g)	Gas volume factor (GVF)	Lockhart-Martinelli parameter (χ)
Gas single phase (wet gas)	>0.7	>0.9	<0.3
Multiphase flow	0.4–0.69	0.5–0.89	0.31–0.61
Liquid single phase	<0.3	<0.5	>0.62

- *Intermittent flow (plugging and slugging)* occurs when gas flows in slugging or intermittently (not continuously), and liquid dominates the total void fraction. Liquids have moderate high velocity and gases have low velocity. This is a more frequent flow before separation, it normally occurs in a black oil system with low GOR.
- *Bubble flow* occurs at a high liquid velocity and is characterized by gas bubbles floating over a continuous liquid phase. Here the multiphase flowmeter (MPFM) works with high accuracy.
- *Annular flow* occurs at high gas and liquid velocities and with a significant viscosity contrast between the gas and the liquid (μ_g/μ_{liq}). The gas flows in the central core of the pipe and liquid flows as a film on pipe walls.
- *Mist flow* occurs when the gas velocity and the density increase, and then the liquid becomes entrained as droplets in the gas flow. A portion of the liquid travels on the pipe's wall, and another portion flows suspended by the gas flow. This flow is predominant in gas condensate and volatile oil systems.

Fluid current homogenization: This condition occurs when the state of the fluid is closer to one single phase because of the mechanical principles. After separation at low pressure and low velocity, fluids can be separated, stratified, or producing in a slug pattern. Some meters, especially turbines, cannot distinguish between gas and oil. Furthermore, at this condition the turndown can change from 1.0 to 100 in minutes. To reduce this effect, mixers are used to blend the gas, oil, and water; therefore, the liquid and gas travel at the same mean velocity ($V_{liq} = V_g$). The slip velocity is 1.0; in this condition, fluid properties of the mixed fluids are a single value of viscosity, density, etc. Homogenization allows turbine meters to measure fluid velocity with high accuracy. Fig. 2.9 shows how a mixer can homogenize several fluids.

Separation system and efficiency: Well-location and gathering center operation use two- and three-phase separation equipment. After the separator outlet, it is expected that the wellhead pressure drops by >60%; under this condition, the flow regime could be a stratified pattern flowing at a laminar flow regime. In this state, the flowmeter measures the flow with acceptable accuracy. Three-phase separators are the most efficient devices to separate water and gas from oil. In non–water production reservoirs, a two–phase separator could be a good choice, but when the condensate or volatile oil flow simultaneously with water, the total liquid production may represent significant amounts of volume (GVF < 0.5). A three–phase separation system could be the best choice to measure oil, water, and gas separately. Generally, after the separation outlet an orifice meter is used to measure the gas rate; at the oil and water outlet, a turbine meter could be the most economical and reliable choice. For a two–phase separation system, an infrared water cut meter combined with a turbine meter is generally the right choice. The net oil can be estimated by the total liquid measured from the liquid meter times the water cut value of the water cut meter. Fig. 2.10 shows the best

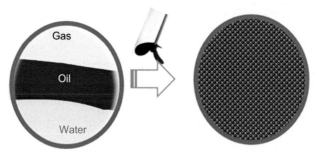

Fig. 2.9 A cross-section of the pipe showing how a stratified flow pattern can be homogenized using a mixing tool.

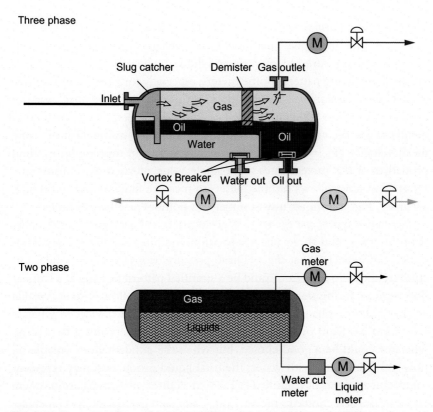

Fig. 2.10 Two- and three-phase separation systems showing possible locations of flowmeters.

location of flowmeters after a separation system. A MPFM is used before the separator inlet.

2.1.3.5 Hybrid Single-Phase Flowmeters (Possible Combinations)

Flowmeters can be used in many possible combinations. Appropriate configurations depend on the properties to be measured (velocity, density, pressure drop, mass flow, and/or chemical element) and the separator design. Falcone et al. (2014) describe three main ways to combine meters: (1) homogenization, test measurement, and separation (techniques depending on homogenization); (2) techniques that measure bulk flow (do not depend on homogenization); and (3) techniques based on flow separation. The flowmeter can be combined as described below:

- *Homogenization, test measurement, and separations*: Except for MPFMs, single-property flowmeters have major issues with measurement under non-homogenized condition; therefore, a mixer device must be used to equalize the flow velocity and estimate the mixture density. The following equipment combinations can be used before a separation system, but this configuration requires human intervention and constant device calibration:

- Two densitometers plus velocity: requires a meter to measure total velocity (turbine), a densitometer to measure mixture density, and second, a densitometer to measure oil phase density at the separation test system. Additionally, oil, water, and gas-phase voids can be estimated with a mathematical procedure. This is a less expensive option but sometimes is not practical. This option cannot measure the gas rate.
- Velocity plus momentum requires one meter to measure the total velocity and an orifice meter (differential pressure) to measure the gas rate. This is also an inexpensive option; however, it cannot measure the phase densities and cannot distinguish oil from water.
- Momentum plus density: Gas rate flow and fluid average density can be obtained, but need another densitometer after the separator to measure oil density. A Venturi meter can measure the fluid momentum ($\rho.v^2$), and a densitometer can measure density of the total fluid. Use a mathematical calculation to get fluid velocity (v).
- Interrogation rays plus mass flow: Using a Coriolis meter, the total flow rate can be estimated with high precision. This meter combined with a neutron interrogation or infrared ray can estimate water cut in the bulk current and therefore quantify the oil and water rate. However, this option does not measure the gas rate, and it can be very expensive and inaccurate if GVF is higher than 0.8.

- *Non-homogenization, measurement, and separations*: generally placed before a separation system. The fluid components flow at different velocities, so the phase velocity must be measured separately as well as volumetric composition and its respective density (no mixture). This scenario requires quite expensive devices and sometimes environmentally sensitive sources, such as neutron interrogation, to calculate the amount of oxygen and hydrogen present in the flow current, in addition to a pulse neutron activation to measure both oxygen and hydrocarbon atomic velocities. Additionally, a γ densitometer is required to measure the mixture density fluid. The MPFM falls under these categories:
  - Bulk flow phases plus phase velocity.
  - Phase density and water cut.

- *Flow after separation system*: This system measures oil and water directly at the tank; gas is measured using an orifice meter. Important measurements are derived from this data including the gas-oil ratio (GOR), the water cut (ratio of water to liquid), and liquid-gas ratio (LGR). Frequently, oil and water density are measured using centrifuge principles, and gas density is measured using correlations.

2.1.3.6 Multiphase Flowmeter

A multiphase flowmeter is an alternative to separators and individual fluid meters is a three-phase meter. MPFM removes the need and expense of a separator and three individual flowmeters and replaces that equipment with a single, but more expensive, single meter. The technology for these types of meters is rapidly progressing, including nuclear and sonic-type meters. As with Coriolis meters, pricing has become reasonable for many applications. An MPFM is generally placed after the wellhead and before a separation system, which allows for better MPFM utilization than placing it after the separation system. To allow gravity to separate the phases and generate uniform bubble flow pattern, the MPFM should be set up in a vertical upflow position. Fig. 2.11 shows the main components of an MPFM, which are listed below:

- Venturi meter measures the total mass flow and bulk velocity by differential pressure.
- Dual-energy γ-ray densitometers or neutron interrogation is a detector to measure phase fraction and mixture density and features. This nonintrusive device directly measures the density of a mixed fluid. It is able to separate the liquid from the gas phase by measuring the atomic velocities of oxygen and hydrogen.
- Infrared meter is an electronic detector that measures the capacitance of water in an oil-dominated flow. Also a nonintrusive device, it estimates the water cut or the amount of fraction of water volume in the total liquid.
- Inductance and capacitance meters are electronic devices that measure water conductivity and resistivity, respectively.
- Gas composition meter estimates the percentage of hydrocarbon composition (e.g., C1, C2, C3 up to $C7^+$) in a gas stream.

A new generation of MPFM can also measure other critical properties, such as solution pH, water salinity, wax content in oil, gas gravity, and oil gravity; estimate the flow regime; and calculate Reynolds, Bond, and gravity numbers, and the solid content in fluids.

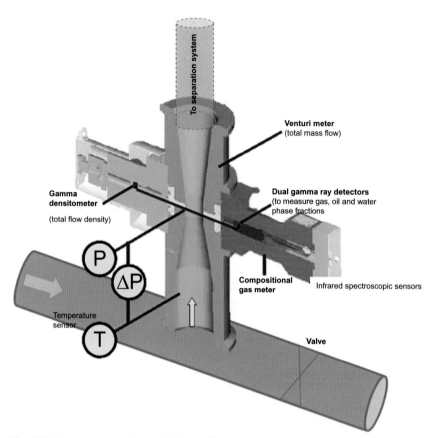

Fig. 2.11 A prototype of a multiphase flowmeters (MPFM) integrated with several devices, such as a γ densitometer, Venturi meter, dual γ-ray detector, infrared ray spectroscopic meter, compositional gas devices, and temperature sensors.

Direct Flow Estimation

Direct estimation methods include compressor volumetric estimation for gas rates and tank level variance for liquids. However, these methods quickly lose their accuracy when multiple wells are combined to feed tanks or compressors. Requires a lot of allocation and calculation processes to improve accuracy.

Virtual Flow Estimation

Where instruments and direct estimation are not possible or economic, virtual estimation can be used. Virtual estimation can be done using real-time nodal analysis or using a data method—like a neural network—based on historical well tests.

For these methods to work requires real-time data readings for pressure and temperatures, so one must install instruments to gather necessary data. Instrumentation is discussed in Chapter 5.

2.1.3.7 Flowmeter Selection

DOF workflows can provide significant value to justify the investment in accurate flow measurement of all phases and sometimes even compositions. Flowmeters should not be selected on cost or price, but on the accuracy needed to support the value-added workflows. Table 2.1 summarizes the factors for selecting flowmeter devices based on: principles of measurement, property to be measured, reservoir fluid most applicable, GOR range of operation, water cut range of operation, accuracy as a function of flow rate, turndown maximum and minimum reading rates, and vendor tolerance.

2.2 CONTROL TECHNOLOGY BY FIELD TYPES

2.2.1 General Control Technologies

The most important control technology for realizing the maximum benefits of the DOF is remote or autonomous means for controlling a well. The best example of this technology is the automatic choke.

An automatic choke lets you remotely adjust the flow in a well—from complete shutdown to restart and nearly every setting in between. This capability can be useful in many operating situations; for example, wells can be curtailed or shut down when problems arise in production facility equipment, then, when the problem is solved, remotely restarted. Many operators are adding these chokes as standard equipment, especially for rapidly declining wells where the chokes can be adjusted remotely as rates decline and wells can be easily set up with intermittent flow to support field-wide capacity management.

Other important automatic technologies are for artificial lifts. Gas-lift valves need to be highly accurate (not on-off) regulatory valves. Electric submersible pumps (ESP) should have adjustable frequency drives, and rod pumps should have pump-off controllers and remote-activated speed adjustments.

This section provides recommendations for use of these technologies for some important types of fields.

Table 2.1 A Summary of the Most Important Factors for Selecting a Flowmeter

Flowmeter	Principles of Measurement	Property Measured	Reservoir Fluid Applicability	GOR (Mscf/Bls)	Water Cut (%)	Accuracy Flow Fate (Mscf/d) 10, 100, 1000, 10,000,.... 1e6	Turndown	Tolerance
Orifice plates	Differential pressure (all intrusive)	Pressure drop and momentum $(\rho \cdot v^2)$	Dry gas Wet gas Gas condensate	>20	<10		5:1	±2% less 4% of total rate (if no water)
Flow nozzle				>100	<5		5:1	±1% less 5% of total rate (if no water)
Venturi tubes				>5	0–100		10:1	
Rotameter				>10	<10		12:1	
Turbine meter	Velocity meter (all intrusive)	Fluid velocity (v)	Black oil w/low GOR	1.0 < GOR < 3.0	0–100		100:1	±0.25% of total rate (if homogenized)
Vortex meter			High GOR Dry/wet gas	>5.0	<30		10:1	±1% of total rate

Continued

Table 2.1 A Summary of the Most Important Factors for Selecting a Flowmeter—cont'd

Flowmeter	Principles of Measurement	Property Measured	Reservoir Fluid Applicability	GOR (Mscf/Bls)	Water Cut (%)	Accuracy Flow Fate (Mscf/d) 10, 100, 1000, 10,000,… 1e6	Turndown	Tolerance
Reciprocal piston Disk Rotary vane	Positive displacement meter (intrusive)	Volume displaced in a cell (flow rate, Q)	Black oil Heavy oil	<3.0	0–100		70:1	
Coriolis meter	Mass flowmeter	Mass flow rate (m_f)	Black oil low GOR Heavy oil	<5.0	0–100		100:1	±0.05% less 0.5% of total rate (if low GOR)
Multiphase meter +Venturi +Ultrasonic +Electromagnetic	Differential pressure Densitometer Flowmeter	Pressure drop Gas particles Water cut	Multiphase flows plus: • Salinity • Density • Flow regime	0.1–1000	0–100		100:1	±0.02% less 0.5% of total (before separation)

2.2.2 Mature Assets

Mature assets are a special challenge for DOF projects because of older oil field technology and lack of basic computer and technology infrastructure (compared with the current standards). These older assets may not support the instrumentation and automation that characterizes DOF systems. However, these assets may have enough hydrocarbon-producing life to warrant the cost of retrofitting them with wellsite controls and automation.

The cost of sensors and infrastructure (for additional discussion, see Chapter 9) is certainly a key obstacle. Wells are instrumented with pressure and temperature sensors to support virtual flowmeters (Chapter 5) and artificial lift unit operations. High-rate wells in large mature fields (such as in the Middle East) can often easily support downhole instrumentation and full SCADA control platforms. For example, the KwIDF projects in Kuwait supported MPFMs in a water injection area (Al-Abbasi et al., 2013).

In remote fields with large well counts but low rates, the trend is to use local control facilities, which operate with solar-charged batteries and have either WiMax or cell phone telecommunications. Furthermore, because most of these mature wells require artificial lift (e.g., rod pump, ESPs, and others), pump manufacturers offer pump-control packages that can be used independently or that can be integrated with SCADA.

2.2.3 Deepwater Platforms and Floating Production Storage and Offloading

These assets typically have high-volume wells and fields with considerable infrastructure from the initial capital deployment. These assets usually operate like a refinery or petrochemical plant with full distributed control systems (DCS) and onboard control rooms. Considerable instrumentation is installed on the surface systems, subsea wellheads, and downhole.

Three evolving trends help these assets apply DOF. First is the application of downhole temperature and pressure instruments used in each well. This technology is becoming more common place as it has become less expensive and more reliable. These instruments—together with nodal analysis—allow users to perform real-time well surveillance and optimization workflows.

One considerable limitation to applying DOF workflows in these fields has been their relative isolation from support groups. Although the operations group is onboard the facility, the engineering and analysis teams are usually onshore. These onshore groups have been limited in their ability to perform real-time workflows because there can be a considerable time lag to get the data from the remote facility. Further, because getting the data to shore is so slow, only limited amounts are moved at reasonable speeds, with the remainder coming on in a "batch" overnight. Many companies are now laying fiber optic cables to alleviate the lag and improve bandwidth. Some locations are even moving toward satellite communications (Fig. 2.13) to improve data communication and asset collaboration. Companies are employing collaborative work environments for production operations onshore to monitor and control in real time (Chapter 8).

Another evolving trend for these assets is the increased power of DCS and historian systems. These systems regularly have the capability to deploy multivariable control and predictive analytics hosted directly on the control platform.

2.2.4 Unconventional Assets

Unconventional assets are uniquely fit for the DOF. These assets are newer fields with many wells that are drilled and produced from centralized pads, so more DOF infrastructure can be applied economically. The well production does not lend itself to traditional modeling and field planning tools, like reservoir models and nodal analysis, but data-driven models can be used appropriately to manage wells. For example, unconventional wells move rapidly through their life cycle, going from natural flow to artificial lift in months. This means unconventional fields have more data available and need more data-driven workflows. As these fields are developed on multi-well pads, a high degree of instrumentation can be centrally installed and the asset can have local control capability. Most equipment has solar panels to drive long-life batteries. They use produced gas to power valves and instruments, and have some local control devices.

Local control devices have become much more powerful and easy to use. Remote terminal units (RTU) with embedded control capability are widely applied. A newer technology, programmable logic controllers (PLC), is quite powerful yet easily programmed and is being used more frequently in unconventional assets. PLCs can now be programmed with graphical-based languages (as opposed to the old ladder logic) and can include advanced capabilities, such as supporting multivariable predictive control of devices without requiring a gateway-connected computer.

Wells in these assets typically have some form of well pad separation and flow measurement as described. They typically include tubing, casing, and flow-line pressures and at least flow-line temperature instruments. It is also very common for these wells to support automatic valves including wellhead chokes. The automatic remote-activated chokes facilitate supervisory control and optimization. Multiple operators use the automated chokes for intermittent well control or for automated curtailment and restart of wells, to balance production with facility constraints, such as when a compressor goes down.

As unconventional wells progress through their life cycles, they move onto artificial lift. Gas lift, plunger lift, rod pumps, and some ESPs are used. All of these are being installed with local-control elements that can be remotely operated. Operators commonly have the ability to set gas lift rates, pump speeds, or pump-off controls from field collaboration centers, without going to the well site. Real-time data-capture technology (e.g., a product from OSI called PI, one of the data historian software available in the market) can be used stand-alone or integrated with SCADA systems to control, analyze, and optimize multiple lift types. As stated above, unconventional assets rely heavily on data-driven workflows because traditional modeling tools do not fully apply. As these are usually high well-count assets, there are large volumes of data to analyze. SCADA and historian systems are now embedding predictive analytics tools into their platforms. Additionally, OSI's PI application framework translates the traditional SCADA/DCS tag-based data model into a virtual one so that users can get right to the data they need with logical names instead of tag names. These systems, with their added security and robustness, give more power to the process control network and operations staff, instead of moving that processing into the enterprise IT architecture.

2.3 DATA GATHERING AND SCADA ARCHITECTURE

2.3.1 Well-Location Data Gathering and Telemetry

The two main ways to gather and communicate data are wired and wireless. Localized well and control facilities, such as offshore platforms, typically use a wired structure. Scattered, large well-count assets, like the unconventional ones, generally used a wireless strategy. Fig. 2.12 shows a well location with an RTU controller, wireless equipment, and an Ethernet switchboard.

The main issue for DOF systems is that larger volumes and increased frequency of data needs to be gathered and transferred to the control location. For medium-sized offshore and land-based assets, these requirements

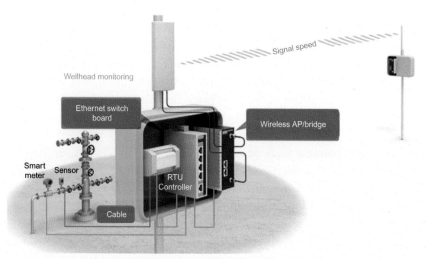

Fig. 2.12 Well-location equipment for data receiver and transmission. *(Image courtesy of Moxa.)*

commonly mean the use of fiber-optic cable to the optimization and collaboration centers. Most of the other means have too much latency or subsampling to enable DOF workflows. Widely scattered assets, like the shale plays, require some wireless technology. Although commercial cell-modem communication is getting more affordable, the variability of plans makes it difficult to fit into a DOF strategy. Therefore, it is most common to see radio-frequency towers used for DOF assets. This area of technology is rapidly evolving so the DOF best practices will be changing very soon.

The IT infrastructure for data gathering is commonly a data historian. This software provides both a data and hardware gateway between the more-secure local control site and the enterprise IT-based DOF systems.

2.3.2 Field Control Devices

While DOF workflows require more data to be communicated up to the enterprise servers and systems, more control and processing power is being pushed to the well pad. It is now very rare to find simple RTU-type units on the well site (as shown in Fig. 2.12). Almost all DOF assets use remote operation centers (ROC) at least. A new version of old technology is the reemergence of PLCs. PLCs have tremendous programming capability and are commonly used in high-frequency manufacturing. These are very robust and stable devices, which are now easily programmed as discussed above (see Section 2.2.4).

2.3.3 SCADA and Distributed Control System

Most modern production platforms and FPSOs use distributed control systems (DCS). Most modern unconventional assets are installing simpler SCADA systems. Many assets are using a generic SCADA system, such as those applied in other industries (e.g., manufacturing). Others are electing to deploy a domain-specific SCADA/control platform—especially on artificial lift wells—or even two systems to meet their needs.

SCADA systems are now in their fourth generation. The first generation was "monolithic," the second generation was "distributed," the third generation is "networked," (as shown in Fig. 2.13) and the fourth generation is the Internet of Things (IoT) generation. Recent SCADA systems are less expensive to maintain while reporting in almost real time. The current recent technology uses open-system protocols and is cloud based. We do not know of any O&G asset currently using a cloud-based SCADA system, but there are many industrial water-conditioning plants using such systems. We think O&G will not be far behind.

Most SCADA or DCS systems with DOF integration technologies are displayed in control rooms to enable collaboration like shown below.

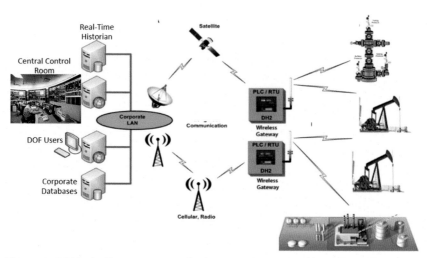

Fig. 2.13 SCADA architecture example showing data transmitted from RTU/PLC well side to a SCADA system.

However, most modern SCADA and DOF platforms need to have some mobile capability as that is becoming critical for all land-based operations.

2.4 SPECIAL NOTE ON CYBERSECURITY

The growth of remote systems that communicate sensitive data from production operations along with the growing occurrences and risks of hacking of communication networks has increased the focus on security for O&G DOF operations and systems. Before the implementation of the digital or connected oil fields, the industry had to face mainly traditional threats, such as natural hazards, human errors, physical attacks to humans and property. Now with the growth of connected sensors, instrumentation, and control systems throughout fields and systems of data connections through cell signals, WiMax, satellites, and cloud servers that contain vital commercial and proprietary data, the O&G industry faces the same cyber threats as any other global industry. Hence, the security of these systems becomes an important aspect of both the O&G industry and the policymakers.

2.4.1 An Overview of Cyber-Attacks in O&G Companies

Like every major company in the world with an IT infrastructure, companies in the O&G industry are fighting against cyber-attacks and their resultant costs. In 2004, 2700 businesses dealing with critical infrastructure had >13 million cybercrime incidents that were estimated to cost them more than $288 million USD and 150,000 h of downtime (Rantala, 2004).

Many major companies in the upstream industry have felt the impact of cyber-attacks including Saudi Aramco, RasGas (Qatar), and Chevron. Traditionally, the goals of cyber intrusions have been to steal intellectual property, business tactics, and information. However, the attack on Saudi Aramco also showed evidence that the aim was to cause physical disruption to the O&G supply chain (Clayton and Segal, 2013).

Cyber threats to O&G companies present a complex and increasingly difficult challenge for the companies, security organizations (governmental and industry), and the public, who may be impacted by such disruptions (e.g., to supply). These attacks are becoming increasingly sophisticated and more difficult to detect, deter and defend against, and the perpetrators are less "lone wolves" and more organized experts of both state-sponsored and self-organized groups. Incidents related to Stuxnet and Shamoom cyber-attacks on energy industry targets appear to be of this nature ((Bronk, 2014; Persons, 2014).

The increasing number of threats and vulnerabilities has warranted industry practitioners, leaders, and policymakers to raise awareness, develop, deploy, and manage solutions to reduce the risk and ensure uninterrupted operations of O&G industry assets.

2.4.2 Cybersecurity Challenges in DOF Systems

DOF vulnerabilities encompass all the vulnerabilities related to IT systems, software, networks, communication, and data systems. The US government Cybersecurity Framework (NIST, 2014) provides a great starting point to understand all aspects of cybersecurity related to DOF systems. The framework has five major steps:
- Identify all threats, actors, and their motivation.
- Set up protective measures against known threats.
- Implement all possible means of detection of vulnerabilities, attacks, and attackers.
- Defend against intrusions by analyzing the sources and their impact.
- After any incident, recover and bring systems back to pristine condition.

Before ubiquitous Internet connectivity, the systems controlling operational processes on the rig, oil field, refinery, or pipeline were closed systems. Now, these control systems operate in an open network environment that makes them vulnerable to potential threats. The need for an open network is required to increase productivity, automated communications from the field to the office or remote control centers, and perform a broad range of electronic transactions (NPC, 2001).

The vulnerabilities in upstream O&G industry include the ones listed below.
- Physical security: Drilling operations are mostly in remote locations, where a large number of resources (equipment, computing systems, sensors, people, etc.) are located. The mere fact that these locations are remote makes them more vulnerable.
- Intelligent sensors and devices: Across the upstream O&G life cycle, a large number of intelligent devices are involved in managing all the aspects of operations (drilling, production, etc.). As evident from some recent incidents, these intelligent devices become an entry point for attacks in the network that can cause disruption in the operation of the O&G field.

- Lifetime of equipment: The lifetime of the equipment, tools with built-in communication channels have a much longer life than IT systems, and as a result can become incompatible with the new IT system leaving holes for attackers to get into the systems and network.
- Machine-to-machine communication: The communication between sensors and devices in control systems is vulnerable to data spoofing that can lead to unpredictable behavior of the device and create a domino effect in an operational environment.
- Communication networks: Many communication channels are now available, including traditional WiFi, Bluetooth, protocols like Zigbee and others. While there are standards on how to use these protocols, there are no industry-wide standards and thus provides a mechanism for someone to exploit gaps in updates in these protocols.
- Traditional Internet Protocol: While the Internet Protocol (IP) has existed for a long time, vulnerabilities like denial of service (DoS) attacks are getting sophisticated, larger, and more frequent, because they are hard for anyone to predict. The increased connectivity of field systems through the IP increases the chances of attacks like DoS.
- Globally distributed stakeholders: Typical O&G field operations use large, diverse teams of company staff, vendors, and contractors who are globally distributed and who have varied degrees of training and experience. Weak communication between stakeholders can lead to bad decisions that could leave vulnerabilities to increases in the threat of insider attacks.

2.4.3 The Actors, Their Motivation, and Kinds of Attacks

Irrespective of known and unknown vulnerabilities, it is good to understand the main category of attackers and the motivation of these actors.

Intellectually curious: These non-malicious attackers take it as a hobby to solve challenges associated with vulnerabilities they discover accidently or from published reports by various cybersecurity industry experts. While such attackers have no malice, their activity could lead to disaster for O&G industry operations.

Former employees: If disgruntled or turned rogue, former employees can sabotage a business based on what they learned about system vulnerabilities while being employees.

Disgruntled insiders: These attackers can be difficult to detect and stop. They exploit the vulnerabilities either for financial gain or to sabotage the business.

Competitors: While it is not prevalent in the O&G industry, O&G is a very competitive industry. Competitor businesses try to gain access to the intellectual property, business secrets, and financial information.

Nation-state actors: With the explosion of organized hacker activities, the various organized groups supported by their country's governing regimes have started to attack O&G industry targets, motivated by many factors.

Terrorist organizations: Their goal is to create collateral damage on a large scale, and an attack on O&G industry falls in that category, because any impact on O&G production can affect millions of people worldwide. The motivation is sabotaging nations, governments, and businesses.

Criminal syndicates: Their goal is to damage in all possible ways, including stealing intellectual property, financial data and information, and even laundering money.

The O&G industry faces many different kinds of attacks. The degree, size, and complexity depend on various factors in digital, smart, or intelligent field systems. The known major threats that industry faces are the following.

Botnet: A botnet is a collection of compromised computer systems, also referred to as zombies that are in full control of a cybercriminal known as botmaster, who is engaged in malicious attacks and more likely unlawful activities. Botnets have been a growing threat as they can significantly affect the operations of O&G industry.

Phishing and Email spamming: This threat is about getting credentials of legitimate system users using deception tactics by the attacker. Typically, a link is embedded in an email or other electronic communication, whereby the link might look very legitimate to the reader. However, when someone clicks the link, it takes him/her to a site the attacker controls and collects the user's information. Later, the hacker uses that information for various other high-level attacks to the systems, networks, or accounts of the legitimate users. For O&G industry, phishing emails are targeted mainly at mid-level managers.

Malware and spyware: Malware and spyware are software, applications, or programs designed to gather information from computer/computing devices without the awareness of the legitimate user of the system. One of the reasons for this threat is the exponential growth of malware signature in cyberspace and the increasing sophistication of malware software. In 2010, Symantec reported >280 million malware signatures compared with just 3 million in 2009. The Shamoom malware incident of August 2012 is one of the most significant cyber-attacks directed against the industry giant Saudi Aramco. Shamoom quickly deleted digital content on hard disks and it is estimated that >30,000 computers systems were affected (Roberts, 2012; Mills, 2013).

Virus: A virus is a program that propagates itself from one computing device to another with the legitimate user's unknowing authorization or intervention. The damage done by a virus is unforeseen and unpredictable as each of the virus is designed for specific activity and purpose and can range from misleading the users to do a certain activity to destroying completely the computing device itself. The virus is embedded in an email, or documents that are shared such as photos and videos. If a virus-infected file is shared through a physical storage drive like a USB flash drive, the virus can spread even without omnipresent connectivity. The threat of viruses for the O&G industry is the same as for other industries.

Worm: Like a virus, a worm moves from one computing device to another and keeps a record of the previous computing environment it was in, thus providing the attackers a trail of information of the systems. These are self-replicating programs unlike viruses, which are fixed but spread due to some human intervention.

Denial of Service (DoS): DoS attack is an incident in which a business's computing systems are unable to fulfill the service requests that are being requested, because its computing resources are overloaded. Typically, this happens when an attacker creates a massive amount of service requests aimed at a particular service with an aim of bringing down the service, such that the computing system both become overloaded in various resources available to it and is unable to take any more legitimate requests. DOS is one of the hardest kinds of attack to predict and prevent.

SCADA attack: As mentioned above, SCADA systems are the heart of O&G industry operations, so its security is crucial. Typically, SCADA transactions are done without close security at the source; thus, the interceptors can read and use for their benefits. In addition, devices in SCADA systems have very limited memory and bandwidth for storing and implementing authentication solutions, so as a result may allow for injection of requests that can create havoc for field operations. SCADA systems have evolved to become a decentralized series of interconnected networks and thus become more vulnerable. If SCADA systems are attacked and infiltrated, the result could be damage to various assets deployed on the

DOF systems, leading to a shutdown or physical disruption, which can result in economic loss, injury or loss of human life, and environmental impact.

2.4.4 Addressing Cybersecurity Challenges

The O&G industry is engaging in multiple tactics to defend against and deter these challenges. For example, automated testing of SCADA protocols uses well-designed tests that can simulate real-world conditions and allows IT professionals to view the limitations and vulnerabilities of devices in the SCADA system and network. From a policy approach, the industry needs to adopt and implement risk assessment methodologies that encompass all systems, interdependencies, and all aspects of cybersecurity. Each company or organization should adopt a comprehensive cybersecurity strategy and framework that is adapted to the operational priorities of the business. More importantly, the O&G industry needs to build industry-wide standards for cybersecurity for digital oil fields, smart fields, and intelligent fields, through cooperation and sharing best practices used in their respective companies. Of course, the industry needs to implement a continuous monitoring of all aspects of vulnerabilities, actors, activities, threats, and leverage emerging technologies like Big Data and artificial intelligence to mitigate risks associated with cybersecurity.

2.4.5 The Future on Cybersecurity

As more instrumentation, local control, and advanced SCADA features are installed in O&G assets of all types, DOF systems must become more advanced and natively include cybersecurity capabilities. As technology advances, prices fall, making it easier to include security at all necessary locations and levels in a DOF system.

It appears the IoT trends will be to place more local control and optimization capability at the well site. Further, the SCADA data is moving to cloud storage. Several companies already offer cloud-based SCADA. This allows SCADA to scale easily both in terms of more field assets and also in terms of more real-time control and optimization. It also becomes more affordable to maintain because IT groups can then "bundle" support of SCADA with the necessary security with all other enterprise systems.

ACKNOWLEDGMENTS

With contributions from Doug Johnson and Satyam Priyadarshy.

REFERENCES

Al-Abbasi, A., Al-Jasmi, A., Goel, H.K., Nasr, H., Cullick, A., Rodriguez, J.A., Carvajal, G.A., et al., 2013. New generation of petroleum workflow automation: philosophy and practice. In: SPE-163812-MS.https://doi.org/10.2118/163812-MS.

Beggs, D.H., Brill, J.A., 1973. A study of two-phase flow in inclined pipe. J. Pet. Technol. 25 (5). https://doi.org/10.2118/4007-PA.

Bronk, C., 2014. Hacks on Gas: Energy, Cybersecurity, and U.S. Defense. Baker Institute for Public Policy. Retrieved from: https://bakerinstitute.org/research/hacks-gas-energy-cybersecurity-and-us-defense/.

Clayton, B., Segal, A., 2013. Addressing Cyber Threats to Oil and Gas Suppliers. Council on Foreign Relations. Retrieved from: www.cfr.org.

Falcone, G., Hewitt, G., Alimonti, C., 2009. Multiphase Flow Metering, first ed. vol. 54 Elsevier, Amsterdam.

Graham, E., 2014. Multiphase and Wet-Gas Flow Measurement. UK-National Measurement System. NEL. Training Course, Aberdeen, UK.

Gray, W., 1978. Vertical Flow Correlation in Gas Well. API Manual. 14BM. Houston.

Mills, E., 2013. Saudi Oil Firm Says 30,000 Computers Hit by a Virus. CNET. http://news.cnet.com/8301-1009_3-57501066-83/saudi-oil-firm-says-30000-computers-hit-by-virus/.

National Petroleum Council, 2001. Securing Oil and Natural Gas Infrastructures in the New Economy. http://energy.gov/oe/downloads/securing-oil-and-natural-gas-infrastructures-new-economy.

NIST, 2014. National Institute of Standards and Technology. Framework for Improving Critical Infrastructure Cybersecurity. http://www.nist.gov/cyberframework/upload/cybersecurityframework021214final.pdf.

Persons, D., 2014. Energy pipeline: cyber-attacks hit oil, gas, just as much as retail. In: The Tribune. http://www.greeleytribune.com/news/business/10355602-113/cyber-oil-attacks-security.

Rantala, R., 2004. Cybercrime against Businesses: Bureau of Justice Statistics Technical Report. US Department of Justice, Washington, DC https://bjs.gov/content/pub/pdf/cb.pdf.

Roberts, J., 2012. Cyber Threats to Energy Security, as Experienced by Saudi Arabia. Platts http://blogs.platts.com/2012/11/27/virus_threats/.

FURTHER READING

Norwegian Society for Oil and Gas Measurement and The Norwegian Society of Chartered Technical and Scientific Professionals, 2005. Handbook of Multiphase Flow Metering, second ed. Oslo, Norway.

CHAPTER THREE

Data Filtering and Conditioning*

Contents

All digital oil field (DOF) systems generate high-frequency data from multiple sensors from most sources in the field. These sensors communicate through the SCADA systems, remote terminal units (RTU), and data historians with multiple corporate data systems as described in Chapter 2.

It does not matter if the workflow is for surveillance, regulation, reporting, optimization, or control; timely and accurate data is absolutely required. However, the requirements for what constitutes "timely" and "accurate" can vary widely for these workflows, depending on their nature and urgency. For example, gas lift flow regulation requires sub-minute frequency and highly precise indication of flow and valve position, while gas lift optimization may require only the valve position and likely needs only hourly or daily indications of gas lift volume and oil production volume. Thus, DOF workflows have differing requirements for high-frequency data and some may use lower frequency data. The data may be acquired at

---

*With contributions from Doug Johnson.

Intelligent Digital Oil and Gas Fields
https://doi.org/10.1016/B978-0-12-804642-5.00003-7

different frequencies from different sources; fit-for-purpose and some high-frequency data may be processed through time-series averaging processes.

Additionally, user trust and acceptance of DOF systems can be severely limited by bad or questionable data. It is imperative to manage effectively the DOF system data, which is especially difficult because of the large number of data sources and databases involved. Solid data management procedures are necessary but cannot keep up with all data values at all times. Engineers and decision makers depend on data having the highest quality; that is, data that has been processed by data validation, filtering, and conditioning procedures. The IT department, SCADA specialists, and instrumentation specialists are often tasked with ensuring the data quality. Rather than counting on each database to perform its own data management appropriate to DOF, it is best to implement a DOF-based data validation and conditioning system, across the entire DOF implementation.

This chapter presents the major features of such a system, which includes: (1) data processing, (2) basic error detection, conditioning, and alerting, (3) well and equipment status detection, (4) advanced validation, and (5) workflow-based conditioning. This chapter is a condensed tutorial on how to validate and condition data appropriately for DOF systems. The process flow of the chapter is summarized in Fig. 3.1, which has the main steps for a DOF data validation and conditioning system. One can also refer to a myriad of specialty material on signal processing (e.g., Vetterli et al., 2014) which is not covered here.

3.1 DOF SYSTEM DATA VALIDATION AND MANAGEMENT

All integrated systems use three primary levels of data: raw instrument data, calculated data, and asset hierarchy data. Examples of instrument data are temperature, pressure, and flow. Calculated data can be allocations or forecast data. Asset hierarchy data includes the organization of wells to routes, or facilities, water injectors to producers, etc.

The first step in a data validation and conditioning (DV&C) management system is to check basic data transfer protocols and individual high-priority data feeds for all data types. There are two main levels at which data are checked.

First, the instrument data should be checked. All critical instruments, for example, temperature, pressure, and flow rates, should undergo a range and freeze check. Since a DOF system typically has hundreds to thousands of data types, also referred to as data tags, this is a large task that should be

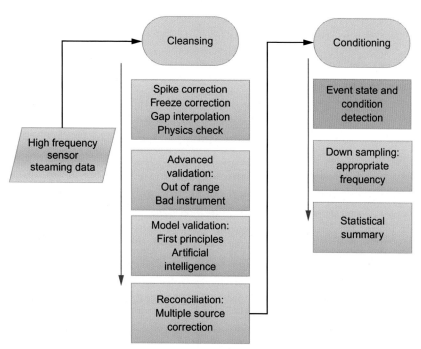

Fig. 3.1 Process flow and tasks to clean, validate, and sample data for DOF workflows.

automated, but there must be some manual checks. It is important that this task be done for the integrated database of all the data (see Chapters 1 and 2), so one is checking the data after it has been through all of the transfer and load processes. Note that it is relatively common for DOF systems to receive polling and job status information from SCADA and other source databases. This information is very helpful but often insufficient to assess the data quality. If the instrument reading is out of range, high or low, or if the data is frozen and has not moved within a tolerance for a considerable time, then something is wrong. It may be the instrument itself or there may be a problem with the transfer, but something is wrong. Most calculated data can be checked this way too.

Second, at the other extreme of data checks is to look at high-level data, instead of individual instruments, which should include both manual and automated data checks. The best practice is to have two ways to look at the data manually: trends and grids (maps). The trends data display should show 7-day aggregated information on route sub-asset, facility, or other organizations. The grid display should visualize all wells for the asset, then the 10 or so most important values for the well. This list should include some non-SCADA information like the producing type. A user can quickly look through these pages every day for a few minutes and find problems quickly.

A special type of manual data check should occur when a well or equipment is first brought into the system or needs some particular analysis. There should be a special query available that lists all data for only that well/equipment. Until a user has manually validated all of this data, the well stays in quarantine and should not be added into the DOF system. The user who does this analysis and makes this decision needs to be a field person with domain expertise. Only when the data is confirmed to be accurate should this well be added into the system.

The automated checks should not rely on poll times or communication checks. The best method is to look at data in ways that conform and test against physical reality. For example, it would be very unlikely that a route would have a rate totalizer that decreased during the day, or that daily volumes should not change during the day. If some quantities fluctuate, that might indicate either direct individual equipment values are changing or that the numbers of wells in a route or other organization are changing, which is not likely. In these situations, automated alerts can be issued.

Once the data is found to be suspect either at an individual or aggregated level, it needs to be flagged to the user or a ticket created for resolution. As soon as one of the above data checks fails, the user interface should indicate the issue, for example, stoplight-type indicators are common. As soon as the data becomes suspect, a yellow color is displayed; after time has elapsed without resolution, then the stoplight turns red. The time between yellow and red is generally determined by the update time required by the fastest workflow. For example, if that is a daily workflow, then it could be 4 to 8 h between a yellow and a red alert.

Finally, each individual reading should be minimally conditioned at this stage, which means two things: first, if the data fails the range or freeze check, then the last good value is held, and second, a small filter is applied to remove high-frequency noise. We recommend this be a digital implementation of an exponential average filter and not a moving average.

3.1.1 Data Processing

Typical surface pressure, temperature, and flow rate sensors capture data values at 1 s or higher intervals. Data are transferred through the system using an RTU or PLC to a SCADA environment, then on to a data historian application. Depending on the storage data duration and storage size, the data historian software is set up to capture data with specific acquisition rates (seconds, minute, hours); internally, the application performs a series of calculations performing the following tasks:

- Transform electronic signal to physical data types (e.g., pressure, temperature, rates, power, friction, choke settings, etc.).
- Convert raw data to specific units (e.g., psia, °F and Mscf/s, or bbl/day).
- Clean and filter bad data, data spikes, out-of-range data, and frozen data.
- Identify and replace missing data with rules, imputation, or reconciliation.
- Down sample raw data in single data points, for example, 60 s in 1 min, 60 min in 1 h, etc., through time series averaging.
- Summarize high-frequency data to statistical values over lower frequencies.
- Aggregate missing data or replace information for misleading data through statistical interpolation.

3.2 BASIC SYSTEM FOR CLEANSING, FILTERING, ALERTING, AND CONDITIONING

In a real-time process, data quality is a common and persistent issue, and it takes time to repair or to replace the information. Typical problems include the following.

- Missing data: Most of the time this occurs when connection is lost between SCADA and RTU systems. This generally happens in hostile environments, with extremely low or high temperatures, or winds at high velocity. The lost signal appears as a gap in the data (null value) or as a "frozen" data point (flat line).
- Data out of range: Occasionally signal errors can occur in remote systems or metering systems, such as non-calibrated meters equipment like orifice gas meters or turbine meters, which result in values that are out of range of the reasonable or acceptable range.
- Frozen data occur when a data value is repeated several times due to malfunction of electrical equipment or no transmission. To avoid null or zero values, the system automatically repeats the previous value. Most of the time this is an issue related to the programming language of PLC or RTU devices.
- Data spike is a natural error signal that is outside the manufacturer's tolerance; it is random, occurs infrequently, and is typically caused by

external factors, such as power interruption or sensor error. When the RTU or transmitter signal is malfunctioning, data spiking can increase. Often, data spikes can be mistaken for gas or oil wells flowing with slugging conditions. The two can be distinguished because data spikes occur with high frequency (every second), whereas the spiking behavior in a slugging well occurs with lower frequency (over several minutes).

- Data do not follow any physical behavior; a signal or multiple signals do not correspond to physical requirements. For example, gas rate is a function of pressure response, so when gas rate increases, by definition pressure should decrease and vice versa. When the well is completed and shut-in (gas rate = 0.0), the tubing pressure builds up in proportion to static reservoir pressure. So if one obtains in case of a reading that does not accurately reflect these known physical conditions, it's probably an error.

The above conditions relate to the sensor signal or SCADA. But engineers must learn to distinguish between these types of problems and data that may actually be alerting them to real issues in the production system.

Fig. 3.2 shows a plot with gas production and surface pressure. The data is logged every minute for a 24-h period. The plot shows several of the data situations described above: data spikes, missing data, gas rate out-of-range without pressure unchanged, frozen gas rate, and sometimes zero, when pressure is similar to the gas rate above zero.

3.2.1 Data Validation System Architecture

The data validation process can be considered part of the overall data quality control (DQC), involving both raw and processed data. Production DQC is defined as all those operational procedures that are being used routinely to ensure the reliability of monitored data. It consists of an examination of data to detect errors, so that data may be corrected, filtered, or deleted. Quality control of raw data is performed to eliminate errors of measuring devices such as sensor malfunction, instability, and interference, to reduce potential corruption of processed data. These procedures should be done as close to the source as the data allows. In conventional practice, a first level of these practices is often done in the historian software. But some companies perform the quality checks once data have been communicated from the historian to a database. This second option often leads to quality issues and uncertainty about data

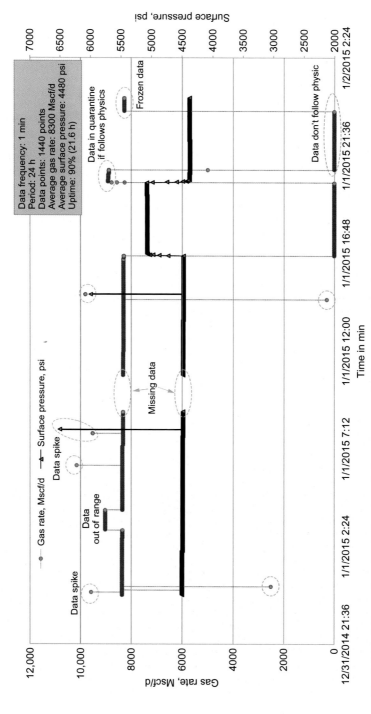

Fig. 3.2 Typical raw data (1-min frequency) showing many data errors including data spike, out-of-range, missing data, frozen data, and meaningless data.

quality. Many modern sensors in SCADA systems have built-in functionality for signal processing at the measurement location, that is, "at the source".

Multiple possible checks are listed below.

3.2.1.1 Rate of Change, Spike Detection, and Value Hold

This algorithm includes a test and a valid value setting procedure to correct data spikes. If the test passes, then the current value is passed along. If the test fails, then the last good available value is held for a specific period and an alert is generated for the engineers. The test takes the current value and a user-entered tolerance to check if the current value has spiked and will set the output value according to the following logic (pseudo-code):

> If $ROC\_test = off$, then skip this test
> $Y_{(t)} = $ If $ABS\big(X_{(t)} - X_{(t-1)}\big) > Tolerance$, then $(error\ and\ set\ alarm)$, else $x_{(t)}$,
>   and set alarm and counter $= 0.0$
> $Y_{(t)} = $ If $Y_{(t)} = error$, then $Y_{(t-1)}$ else $Y_{(t)}$, and increment counter
>   If counter $>$ user limit reset $Y_{(t)} = error$
>   t is time stamp

3.2.1.2 Out-Of-Range Detection and Value Clip

This algorithm includes a test and a valid value setting procedure. If the test passes, then the current value is passed along. If the test fails, the current value is reset to a valid higher or lower value. This test does not assume the tag has gone bad but that the reading has gone out of range. That is why the value is not set to bad but reset to a value clipped to the nearest maximum or minimum of the acceptable range limits. The over-range (OFR) test should catch an instrument failure due to an over-range value failure, because a failure should cause an immediate spike. The test takes the current value and a user-entered minimum, maximum, high-clip, and low-clip value. The logic is:

> If $OFR\_test = off$, then skip this test
> $Y_{(t)} = X_{(t)} > Max$ then clip above value and set alarm, else $X_{(t)}$ and set alarm $= 0$
> $Y_{(t)} = X_{(t)} < Min$ then clip below value and set alarm else $X_{(t)}$ and set alarm $= 0$

3.2.1.3 Freeze Detection and Value Hold

This algorithm includes a test and a valid value setting procedure. If the test passes, then the current value is passed along. If the test fails, the last good value is held for a specific number of cycles. The test takes the current value and a user-entered tolerance and window to check if the current value has frozen and will set the output value according to the following logic:

> *If FRZ_test = off , then skip this test*
> $Y_{(t)} =$ *If* $ABS\left(X_{(t)} - X_{(t-1)}\right) >$ *tolerance for " time window" , then*
> *(error and set alarm), else* $X_{(t)}$ *and set alarm and counter* $= 0.0$
> $Y_{(t)} =$ *If* $Y_{(t)} =$ *error, then* $Y_{(t-1)}$ *else* $Y_{(t)}$, *and increment counter*
> *If counter > user limit reset* $Y_{(t)} =$ *error*

3.2.1.4 Statistical Detection and Value Hold

This algorithm includes a test and a valid value setting procedure. This test is used to detect a fast-cycle instrument drift. The first step is to calculate the signal's mean and standard deviation (σ) over a time window in the past. The test algorithm checks if the current signal value exceeds the mean ± 3 times the standard deviation. If the test passes, then the current value is passed along. If the test fails, then the last good value is held for a specific number of cycles. The test takes the current value and a user-entered window for statistical calculations. The calculation logic for statistical process control (SPC):

> *If SPC_test = off , then skip this test*
> $Mean_X = Mean(x, time\ window)$ *and* $\sigma_x = \sigma(x, time , window)$
> $Y_{(t)} =$ *If* $X_{(t)} > Mean_x + 3^*\sigma_x$ *then error and set alarm, else* $x_{(t)}$, *and set alarm* $= 0.0$
> $Y_{(t)} =$ *If* $X_{(t)} < Mean_x - 3^*\sigma_x$ *then error and set alarm, else* $x_{(t)}$, *and set alarm* $= 0.0$
> $Y_{(t)} =$ *If* $Y_{(t)} =$ *error, then* $Y_{(t-1)}$ *else* $Y_{(t)}$, *and increment counter*
> *If counter > user limit reset* $Y_{(t)} =$ *error*

3.2.1.5 Filtering

The data conditioning system uses a simple exponential moving average or low-pass filter. The filter takes a filter constant, in time, the current data reading, and the last filtered result to calculate the current filtered result according to the following algorithm:

> $Y_{(t)} = 1 - C_f \times Y_{(t-1)} + C_f \times X$ *and*
> $C_f = 1 - Exp^{(t_s/a)}$

where t_s is the time window and a is the user entered filter time constant.

Other averaging schemes or filter transforms may be used by operators.

3.2.2 Advanced Validation Techniques

The basic validation described above is meant to keep bad data (regardless of cause) from corrupting DOF databases. Advanced data validation is meant to detect particular problems with data. The problems could be bad meters, poor operation of surface facilities, abnormal operating conditions, etc. These techniques make considerable use of fundamental production attributes, statistics, and model-based methods.

After basic single-variable validation, it is common to look at multivariable calculations with respect to fundamental laws of production. As the use of flow rate meters is increasing dramatically in many assets, it is crucial that these meters are calibrated for precision and accuracy. Just because they read a number and are not frozen does not mean the results are good. Operators spend many hours chasing suspect meter data. Many automated methods can be used to check meter accuracy as well. For example, mass balances are commonly used in calculations as checks. In addition, volume or flow ratios, like gas–oil ratio or water cuts, should not shift drastically in a short time. If they do, it is likely a metering issue. A metering issue could be due to the meter itself needing calibration or it could be due to the separation equipment not operating correctly; for example, water gets carried over with the oil channel and is measured as oil. Again, these checks determine that a problem exists and approximately where it is but may not directly identify the specific cause. The ratios described here necessarily use data for the well status required. For example, these tests should not use data when the well is down or not at steady state.

Another common practice is to use SPC practices. There are two reasons to perform these checks: to detect a shift in (a) the process itself or (b) bad data. Although it is beyond the scope of this book to present SPC, many texts explain it. The basic rules make use of a long-term average or mean and the signal standard deviation. For example, you can calculate a 7- to 30-day moving average and standard deviation. Note that these calculations should use the status signals described above (the average and standard deviation should only use data from when the well is up or in steady state). The rules for abnormal process behavior are applied under these conditions:

- if the current data (daily average) is beyond the average ± 3 standard deviations;
- if the last 2 daily averages are beyond the long-term average ± 2 standard deviations;

- if the last 3 daily averages are beyond the long-term average ± a single standard deviation; and
- or if the last 5–7 daily averages on one side of the long-term average or the other.

These rules are based on the statistical fact that less than 5% of normal process variation should be beyond 3-day standard deviations. As the well behavior does move over time, these calculations need to be monitored and may need to be changed over time as the well changes. Furthermore, any time a large variation occurs—for example, 10–15 times standard deviation—then the data is likely bad.

3.2.3 Model-Based Validation Methods

The final type of the technology used to detect bad data is model-based methods, which include two types: first principles or artificial intelligence (AI). Any component that can be modeled with a first-principles technology can be set up to have a value predicted from the model that is also measured for comparison. Then you can look for deviations between the model-based value and the actual reading from the instrument. Of course, deviations could be due to either a model issue or a measurement issue. Further analysis is necessary to determine the exact cause. AI can also be used to detect bad data or process changes that may be due to bad data. Self-organized neural networks or k-means clustering can be used as fault detectors. Furthermore, they can be used with very large amounts of data. The AI methodology is to download data and cleanse all bad data for the data set. Then train the cluster on the good data. When the cluster algorithm is implemented in real time, it issues an alarm if any data pattern is observed to be outside of the trained cluster.

3.2.4 Data Replacement Techniques

If data validation routines detect bad or missing data, values may need to be "created" to fill the gaps. For DOF, it is imperative to work with continuous (and often high-frequency) data; most of the automated engineering workflows (Chapter 5) work with continuous data, for example, pressure and rates. In cases for which pressure or rates are missing or marked as bad data, the workflow should replace or populate this unreadable data with alternative available source data. In our experience, we believe there are several levels of complexity to replace or fill in data, for example:

- Level 1: Simple averaging (summarizing).
- Level 2: Extrapolation, data follows trend and tendencies.
- Level 3: Replace by data-driven analytics, for example, artificial intelligent components (Chapter 4).
- Level 4: Replace data through physics-based calculation. Using an engineering model-based physics to replace pressure in build-up time or to replace flow rates data using virtual metering system (covered in Chapters 5 and 6).

Our experience in several projects is that replacing data with models often presents challenges (Al-Jasmi et al., 2013a,b; Rebeschini et al., 2013). For example, in one water flood that had multiphase flowmeters (MPFMs) and real-time artificial lift data, we used different processes to aggregate data. These included artificial neural networks (ANN), fuzzy logic (FzL), well-performance evaluation (analytic flow analysis), a one-dimensional (1D) analytical model, and 3D numerical models. We learned that there is not a single, reliable, and confident data aggregation process for all cases. A fit-for-purpose model should be determined for each situation and can depend on the reservoir drive mechanism, state of flowing condition and flow regime, artificial lift type, and fluid types.

In the water flood example, we found that using ANN to estimate short-term production was effective but that predicting water cut or oil rate after water breakthrough could introduce errors. The best approach is to train the ANN with a physical model. A 3D physical model cannot be used to replace real-time data directly (month vs. minutes), but engineers can understand that training the ANN to predict the water breakthrough depends on historical data and reservoir properties. The most important factors are:

- Wells producing below or above dew or bubble point pressures.
- Water from water flooding or an aquifer breaking through into the wells.
- Reservoir flow regimes (radial, bilinear, transitional, boundary dominated).
- Flow conditions (steady, pseudo-steady, and -unsteady states).
- Well loading up or well flowing in critical conditions.

Fig. 3.3 shows eight different cases to replace or fill in real-time data. Pressure, gas, or oil rates and water cut are shown versus a period of 7 days. The replaced data is shown by dotted lines. The plot describes the following:

- Fig 3.3A shows a well producing under pseudo-steady-state condition. The pressure depletes (dp/dt = constant), the water rate is near constant, and the total gas rate declines with time. The rate data gap could be reasonably replaced by using simple statistic (averaging).

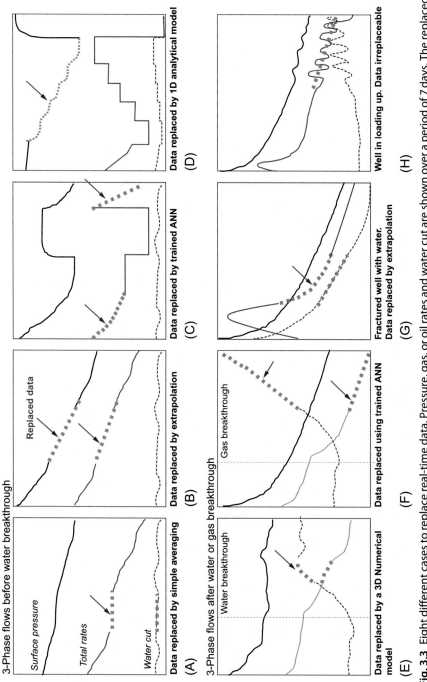

Fig. 3.3 Eight different cases to replace real-time data. Pressure, gas, or oil rates and water cut are shown over a period of 7 days. The replaced data is shown in dotted lines. There are cases with two- and three-phase flows.

- Fig 3.3B shows a well in unsteady state condition with constant water rate. The data is replaced by an extrapolation/interpolation approach.
- Fig 3.3C shows a well response from changing the choke size with constant water cut. The gas-rate data can be replaced by using a predictor such as a trained ANN, FzL, or nodal analysis, before and after changing the choke size.
- Fig 3.3D shows a well during a test with different pump frequencies, from low to high frequencies, increasing oil rate, and decreasing flowing bottom-hole pressure (fBHP). In this case, the fBHP is missing, so a 1D analytical model calibrated with reservoir properties can replace the fBHP data.
- Fig 3.3E shows a well with declining oil rate. fBHP is approximately constant because water injection maintains the reservoir pressure. Water breakthrough increases water cut from 20% to 40%; so in this situation, a 3D numerical model could be the best tool to replace data. An ANN can be trained with a 3D numerical model to predict water and oil rate data instantaneously, as observed in Fig. 3.7F.
- Fig 3.3G shows an unconventional well (low permeability) fractured with slick water, as observed water produces first at high volume, and gas increases and then decreases when a boundary condition is hit. Gas rate and water could be replaced between flow regimes with rate transient analysis (RTA) type curves.
- Fig 3.3H shows a scenario where a well produces under critical condition or water loading up; in this case replacing data is a challenge and might not be practical or appropriate.

3.2.5 Data Reconciliation

Production reconciliation is a data process and statistical method to calculate a final production value when two or more different sources and measurements are available. In DOF systems, it is common that two or three meters can easily mismatch and that not all devices or methods measure production correctly at all times. Quite often, the dispatcher and receiver have a significant discrepancy in fluid readings. Owing to these conditions, the DOF systems require reconciliation methods to correct input data and generate unique readable, cleaned-up, and validated output information. Reconciliation is used to match fluids (e.g., gas, water, and oil) but rarely pressure. Like any other statistical method, the reconciliation process can match only the data within the preset limit, tolerance, and uncertainty values. The limits of reconciliation are that it can only solve mismatches when there is pressure fluctuation, mass or density variations, water cut changes, temperature change, etc. Reconciliation cannot be used when an abnormal situation occurs, such as the pipeline leakage, pump wear, pump blockage, equipment failure, bottleneck in pipeline, etc., which must be attended to immediately.

3.2.5.1 Reconciliation Method: Example

An oil well is equipped with a Coriolis meter measuring in real-time liquid rate (Well C). The Coriolis meter has a tolerance of 10%; the average liquid rate in Well C is 850 bbl/d; the total uncertainty is ±85 bbl/d. Well B is producing 1300 bbl/d using a trained virtual meter. The production engineer mentions that the confidence interval is 6% because physics-based calibration was used to calibrate the virtual metering; therefore, the total uncertainty is ±78 bbl/d. Both wells are measured to a portal-mobile MPFM, which has a tolerance of 3%. The total sum of independent points is 2150 bbl/d, but the MPFM measures 2265 bbl/d (+115 bbl/d above the manual sum). Fig. 3.4 describes the wells.

The method describes:

Uncertainty calculation (U_t): $U_t = \gamma \times (1 - \% \text{Tol})$.

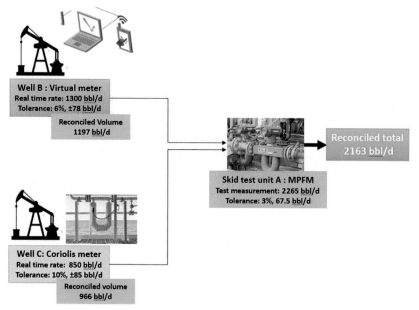

Fig. 3.4 Data reconciliation method with three points of measurement: coriolis, virtual, and multiphase flowmeters.

Standard deviation (σ): $\sigma = U_t/2$

Initial balance (Bal): A–B–C$=0$

Yield between B and C: $Y=B/C$

Penalty of measurement (P_m): $P_m = \left[\dfrac{(y_i^* - y_i)}{\sigma}\right]^2$

Objective function ($f_{(y)}$) minimization of the total penalties:

$$f_y = \sum_{i=1}^{n} \left[\frac{(y_i^* - y_i)}{\sigma}\right]^2 = 0.0$$

where

- y_i is the initial measurement at point A, B, or C using the flow rate tool, either device or calculations.
- y_i^* is the measurement at point A, B, or C after reconciliation.
- %Tol is the manufacture tolerance in percentage.
- U_t is the uncertainty calculation of measurement
- Y is the yield between points B and C
- σ is the standard deviation calculated as the half value of the uncertainty.
- Bal is the initial balance in the total sum of the independent measurements, which should be 0.0. If the total sum is different than 0.0, then we need reconciliation.
- P_m is the penalty function of the measurement or device.
- $f_{(y)}$ is the total objective function, which should be minimized to 0.0 if condition of Bal$=0.0$.

Table 3.1 shows the final values after reconciliation. Note that Coriolis has been affected by $+116\,\text{bbl/d}$ (14% error), virtual meter by $-103\,\text{bbl/d}$ (8% error),

Table 3.1 Reconciliation Method Among a Coriolis Meter Setup in Well C, A Virtual Meter Setup in Well B, and the Total Fluid Measured in Point A With An MPFM (Note That There is a Mismatch Between the MPFM and Other Meters)

Well	Flowmeter	Measured Flow Rate (Bls/d)	Uncertainty U_t (95% Confidence)	Standard Deviation	Reconciled Value (Bls/d)	Penalty, PM	Measurement Error (%)
A	MPFM	2265	68	34.0	2163	8.9	−4
B	Virtual meter	1300	78	39.0	1197	6.9	−8
C	Coriolis meter	850	85	42.5	966	7.5	+14
B+C		2150	Objective function—minimize penalty➜			23.3	
Imbalance (A–B–C)		−115	Imbalance after reconciliation➜			0.0	
Yield $Y=B/C$		57			New yield %	55.3	

whereas MPFM is affected by −102 bbl/d (4% error). The next step is to review with the production engineer why virtual metering has a mismatch of 8% and acceptable error should be below 5%. Additionally, the Coriolis meter needs to be calibrated or checked for the mass and density of the fluid.

We recommend the use of reconciliation processes for daily production monitoring and equipment surveillance. Table 3.1 shows how a Coriolis meter can be uncalibrated and a virtual meter untrained, which results in data mismatch relative to test measurement. The ultimate impact would be misallocation of the produced fluids from the sales tank back to the wells.

3.3 CONDITIONING

After the data have been validated and replaced, if necessary, they can be conditioned appropriately for specific workflows. There are two types of conditioning used in data applications for streaming real-time data: noise filters and statistical calculations for use in specific workflows. Implementation of these techniques is the key to keep workflows from becoming overwhelmed with too much data. It is best to custom-fit the conditioning methods to the specific workflows of interest. Most SCADA systems, historians, or databases have statistical tools to condition data. Three types of conditioning are described below, which apply to most DOF projects: down sampling from the high-frequency data; summation into daily, monthly, and other specific times; and special status indicators appropriate for DOF workflows.

3.3.1 The Level of Rate Acquisition (Data Frequency)

Pressure, temperature, and flow rates data can be acquired every second or more if data storage is available. Often that level of data sampling is not required, as discussed below. Production, completion, geologist, and reservoir engineers are the ultimate end users of the data and must decide on the requirements. Houze et al. (2017) describe how data can be classified into low and high frequencies (as described in the points below). However, we classify data depending on the acquisition rate, with a modified summary presented in Table 3.2.

- *Low-frequency data*. Gas, water, and oil rates can be taken daily and pressure can be taken an average of 24 h; thereafter, the data can be summarized to weekly, monthly, quarterly, and annually. This category includes well test processes that measure to the separator and tanks the

Table 3.2 Raw Data Frequency and Down Sampling Data for Production and Reservoir Engineering Studies

Raw Data	Natural Data Frequency	Down Sampling and Summarizing	Engineer Study and Evaluation
CHP, THP, FLP Q_g, Q_o, Q_w, DTS	Every second for orifice, Venturi, and temperature-pressure gauges, etc.	Few seconds	Real-time production and pump monitoring; flow profiles
	Few seconds for turbine, vortex meters	1 min	Generate warning message if pump values are deviated
	Few minutes for Coriolis and MPF meters	1 h	Generate alarm if production issues persist for >1 h
		1 day	Well performance (nodal analysis), RTA, DCA, etc.
		1 week	
		1 month	DCA, MBE, 3D numerical model, economic analysis
		1 year	Financial calculations
Downhole BHP-BHT	Deciseconds	No down sampling 1 h	PTA RTA
Downhole DAS	Deciseconds		Fracture diagnostics; proppant placement; multiphase flow in tubing

With contributions from Doug Johnson.

total rates of the day and are used for allocation and decline curve analysis (DCA). Wellhead pressure is taken before and at the end of the test.

- *Intermediate frequency data.* The pressure, temperature, gas, oil, water rate data are taken every hour and then summarized to daily averages. In this category, the hourly to daily formats are suitable to perform RTA, allocation, and early time DCA, which depends strongly on independent measurement of oil, water, and gas per hour. In at intermediate frequencies, the data relies rely on constant calibration of pressure gauges and flowmeters.

- *High-frequency data.* The gas, water, and oil rates can be taken every second for real-time and DOF operations. Some meters need more than 15 s to measure the rate, such as Coriolis and MPFM. Orifice and venturi meters can measure the fluid in seconds or continuously. Vortex and turbine meters depend on the velocity of spinners, generally in a few seconds. Downhole memory pressure sensors can take the pressure signal every decisecond (1/10 s); this high-frequency rate is needed to identify the early time properties in a pressure transient analysis (PTA) test, such as the wellbore storage coefficient, skin factors, etc. Others are fiber optic distributed temperature sensing (DTS) systems.
- *High-definition data.* Data that stream at terabytes per minute, such as fiber optic distributed acoustic sensing (DAS) and wellbore acoustics/μ-seismic data.

Fig. 3.5 depicts the data storage for a real-time database, a 24-h production test database, and an extra-large database designed for seismic and fiber optic information. All the data are stored and compressed in a master data bank. A second step is used to clean up, detect spikes, filter and condition data and restructure data in a different SQL data table. Depending on the final engineering purpose, the data will be organized and downsampled in seconds, minutes, hours, days, and months and then summarized depending on the final utilization. For example, PTA data tables are stored in deciseconds, RTA in hours, DCA in days, well test performance (nodal analysis) in hours to days, and numerical model, material balance equation (MBE), economical analysis and financial calculation in days to months.

3.3.2 Down Sampling Raw Data

The data is frequently reduced, filtered, or simply downsampled to manage the data for engineering purposes. In signal processing, this is called "decimation" and is commonly used for PTA or RTA to reduce the pressure signal data by 10:1 or so versus the raw data. Down sampling is a technique of data processing that reduces the data frequency from seconds to minutes to hours while preserving the main signal changes, variations, and physical meaning of the data. Fig. 3.6 shows an illustration of the data downsampling process, for an example RTA analysis.; note that data in seconds are downsampled to hours. Fig. 3A is a plot of the 1-s data taken over a 24-h period; Fig. 3B illustrates the data being both cleansed of out-of-range and spikes and downsampled; and Fig. 3C shows the downsampled data at 1 h intervals for the day.

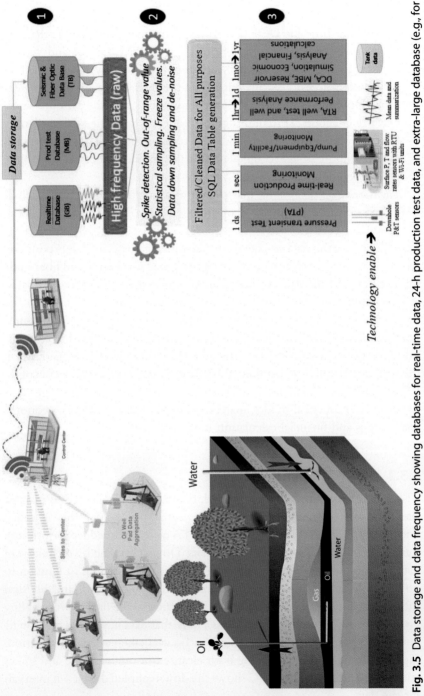

Fig. 3.5 Data storage and data frequency showing databases for real-time data, 24-h production test data, and extra-large database (e.g., for seismic data). All high-frequency data are stored and compressed in a master data bank. A second step is shown to reorganize the data in different SQL data tables for different engineering analyses or workflows.

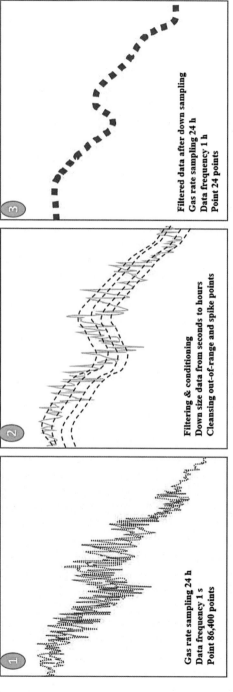

Fig. 3.6 Illustration of data down sampling. (A) Gas rate in 24 h with high-frequency data shows many data spikes and out-of-range values; (B) filter and condition data and then reduce to be sampled at 1 h per day (C).

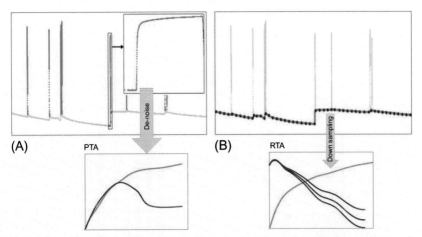

Fig. 3.7 Examples of a surface pressure response for 6 days showing data utilization for PTA (A) and RTA (B) analyses. Note that for PTA pressure, the surface pressure spikes are essential for build-up analysis whereas for RTA, these data spikes should be cleaned up.

Down sampling strongly depends on the final user's purpose, that is, how the data will be used. When the data are used for PTA, down sampling is not required. When the wells are shut in, the casing head pressure (CHP) signal responds like a spiked signal; in this situation, the algorithm should be smart enough to capture the signals when the well is shut-in or a physical event occurs in the well. Fig. 3.7 uses the same high-frequency data (every second) for a period of 7 days. The oil well had six unexpected flow interruptions, including when the well was shut-in during that time. CHP builds up generating important data for PTA, and the pressure peaks should not be cleaned up and should be stored in the database as raw data. The same information could be used for RTA in this situation, that is, flowing CHP or BHP (fluid rates >0.0) should be cleaned up and filtered of those pressures spikes. Fig 3.7A shows the real-time surface pressures during a shut-in time for PTA and Fig 3.7B shows the same data filtered for RTA evaluation.

3.3.3 Summarizing From Raw Data

Summary calculations based on statistics (average, mean, and standard deviation) are used to convert high-frequency data to lower frequencies, such as hourly, daily, and monthly average data. In statistics, we usually calculate the simple arithmetic mean, the statistical dispersion of the data using standard deviation, and shape of the tendency using kurtosis. The users commonly simply sum rate over 24 h and divide by 24. This method can introduce many errors because spikes and frozen data are included in the average. Calculating statistics on the corrected data, the result smooths out higher frequency noise and thus provides better estimates within the summary time windows. Fig. 3.8 shows an example of gas and oil rate for a full year that is measured at custody transfer and tank, respectively. The top chart (A) shows that the gas and oil rates are measured every hour. The total cumulative oil from using flowmeters and applying a shrinkage factor is 135,455 bbl, compared with the total oil sold at dispatch of 132,401 bbl, for an error of 2%. The middle plot (B) shows gas and oil rates summarized per day using statistics; the total cumulative oil is 130,000 bbl compared with the tank which is 132,400 bbl, for an error of 1.8%. The bottom plot (C) shows gas and oil rates summarized per month; note that the total oil error is 3% and the cumulative gas could be up to 10%. The conclusion: data summarized using proper statistics can be more accurate compared with real-time measurement at custody transfer.

3.3.4 Well and Equipment Status Detection Required for Sampling

After the basic data validation is completed and before data are used for any calculations (like daily averages), it is critical to detect well and equipment states. The most basic of these states is if the well or equipment is online or not. Down well data should not be included in data averages for engineering workflows. Examples of states that need to be captured are:
- Well up/down
- Well on test
- Well in high-/low-pressure state
- Well intermittent flow
- Well on lift
- Flare on pilot or flaring
- Compressors/pump up/down

Again the purpose of these states is to provide a downtime status or other equipment state for calculations and conditioning algorithms. With this information, only data relevant to each workflow will be sent to that workflow. Downtime coding only needs downtime data and gas lift optimization only requires data when the well is flowing and on gas lift. Further, well optimization can disregard times when the well may be curtailed due to facility constraints.

The well up/down status likely needs special attention. Wells can shut down fairly fast and valve positions, rates, or pressures can detect this shutdown. When a well starts, it is a dynamic process and involves several

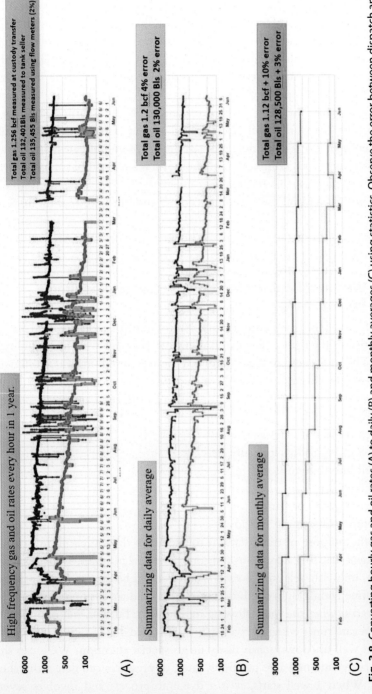

High frequency gas and oil rates every hour in 1 year.

Total gas 1.256 bcf measured at custody transfer
Total oil 132,401Bls measured to tank seller
Total oil 135,455 Bls measured using flow meters (2%)

(A)

Summarizing data for daily average

Total gas 1.2 bcf 4% error
Total oil 130,000 Bls 2% error

(B)

Summarizing data for monthly average

Total gas 1.12 bcf + 10% error
Total oil 128,500 Bls + 3% error

(C)

Fig. 3.8 Converting hourly gas and oil rates (A) to daily (B) and monthly averages (C) using statistics. Observe the error between dispatch and tank data compared with real-time measurements (A and B). Also, compare the error from meter data versus sales data.

production periods or status conditions, each of which may need to be detected; the most important periods or status conditions are "flowing" and "steady state." Other status conditions may be tracked, but these are the most important. Flowing simply means that the well is open for flow. Steady state always takes some time to develop after the well is flowing. There may be a ramping of the choke, lift has to be established, and the "flush flow" has to subside—or maybe the well is loaded and does not immediately flow. In any case, it typically takes some time before the well is truly "up." Many workflows should only use the steady-state data, so it is important to determine when this occurs after startup.

3.4 CONCLUSIONS

"It is all about the data!" We have heard this mantra throughout the industry from managers and engineers on all types of projects and fields. They recognize that the quality of their decisions is only as good as the data quality, consistency, and at the required frequency for the analyses and decisions they need to make. Data problems are inherent in any system and with the advent of DOF high-frequency data, these issues can be more problematic. This chapter presents a primer on how to manage sensor data streams and to recognize data issues from the instruments, and how to treat that raw data to generate quality, validated data. All of the techniques can be deployed "inline" and in real time and do not need to be run in batch mode after the data are collected. The following chapters show how to use data to make engineering decisions in production and reservoir workflows.

REFERENCES

Al-Jasmi, A., Nasr, H., Goel, H.K., Querales, M., Rebeschini, J., Villamizar, M.A., Carvajal, G.A., et al., 2013a. Short-Term Production Prediction in Real Time Using Intelligent Techniques. SPE-164813-MS. https://doi.org/10.2118/164813-MS.

Al-Jasmi, A., Nasr, H., Goel, H.K., Cullick, A.S., Villamizar, M., Velasquez, G., Rodriguez, J.A., Querales, M.M., Moricca, G., Carvajal, G.A., et al., 2013b. A Real-Time Automated "Smart Flow" to Prioritize, Validate and Model Production Well Testing. SPE-163813-MS. https://doi.org/10.2118/163813-MS.

Houze, O., Viturat, D., Ole, S. et al., 2017. Dynamic Data Analysis: The Theory and Practice of Pressure Transient, Production Analysis, Well Performance Analysis, Production Logging and the use of Permanent Downhole Gauge data. Paris, France V5.12.

Rebeschini, J., Querales, M., Carvajal, G., Villamizar, M., Md Adnan, F., Rodriguez, J., et al., 2013. Building Neural-Network-Based Models Using Nodal and Time-Series Analysis for Short-Term Production Forecasting. SPE-167393-MS. https://doi.org/10.2118/167393-MS.

Vetterli, M., Kovacevic, J., Goyal, V., 2014. Foundations of Signal Processing. Cambridge University Press, Cambridge, UK.

CHAPTER FOUR

Components of Artificial Intelligence and Data Analytics

Contents

4.1 INTRODUCTION

As asset yields become harder to assess, extract, and forecast, oil and gas operating companies and service providers must enable real-time decision-making to better predict business outcomes that drive higher efficiencies and utilization to achieve improved bottom-line results and profitability. With the continued worldwide expansion of the digital oil field (DOF), the exploration and production (E&P) industry is rapidly becoming an information- and data-driven business.

If we accept the prediction that the DOF market will exceed $30 billion by 2020 (Markets and Markets, 2015) along with exponential growth in volume and complexity of acquired data, the E&P industry needs to rapidly adopt the new generation of digital transformation, technology, and processes that include the following:

- Implementation of large-scale, Big Data-driven advanced analytics, integrated into role-centric, relevant time workflows.
- Delivery of holistic ability for capture, classification, integration, and interpretation of all the relevant and disparate data sources (geological, engineering, production, equipment, performance, etc.), regardless of the origin or structure.
- The ability to understand advanced analytical trends and correlation models to quickly and efficiently unlock the "hidden" knowledge from all data sets—from small data to large scale and complex data as well as from historic repositories and databases or from fast streaming data.

DOF systems have been used in the E&P industry for several decades and have been commonly known for delivering on the promise of getting the right data, to the right users, and at the right time, for effective asset decision-making, maximized recovery, and improved operational efficiency. However, the expansion of Big Data, evolution of the Internet of things (IoT) and integration of intelligent, virtual sensors requires rapid transformation to an evolving concept of data-driven DOF systems. These trends introduce challenges in the areas of DOF system architecture, data architecture, and data analytics and invite the following questions, which this chapter aims to answer:

- What data architecture is needed in the data-driven DOF to accommodate the ever-increasing demand to leverage the real-time sensor data, that is, the IoT, across the asset?
- If real-time analytics are a must, what are the challenges related to the quality of sensor data and its integration with the historical data for closed-loop analytics and how do we overcome them?
- How will disparate streaming data and even unstructured data, regardless of structure and origin, be integrated and analyzed and how will it be used in real-time automation systems monitor and act?
- What place does data analytics have in the DOF and how can data-driven models be seamlessly positioned and integrated with the physics models?

Data integration is the first step to use data generated by various sources. It is the common notion that the engineers spend up to 70% of their time searching for data, performing data QA/QC, and reformatting data for analytics and modeling routines. Moreover, engineers spend anywhere between 10 and 20 working days manually collecting performance data for annual reservoir performance reviews. The complexity of data types by activity, incumbent in data-driven DOF operations is captured in Table 4.1.

In a DOF environment, data integration must occur quickly to generate value, but this requirement creates challenges as data volumes grow and vary

Table 4.1 Classification of Data Types by Activity, Incumbent in Data-Driven DOF Operations

Activity	Data Types
Production optimization	Production data, real-time data (pressure, temperature, choke settings, gas injection flow, pump parameters), well models
Well intervention	Production data, well logs, geological maps, down-hole surveys, well files and PVT (pressure, volume, temperature) data
Field development planning	Production data, time-lapse seismic and coring data, down-hole measurements, geological maps, well logs and tests, and PVT data
Artificial lift	Pump parameters in real time, pump type, and configuration
Processing of multi-format data	Seismic data (SEG-Y and SEG-Z formats) and well log data (LIS and LAS formats)
Acquisition of fiber optic sensing (FOS) data	Permanent reservoir monitoring, distributed acoustic sensing (DAS), distributed temperature sensing (DTS)
Acquisition of conventional data	Seismic, reservoir, log, flow, completion, relational, intervention, lab, reservoir, and production
Acquisition of streaming data	DTS/DAS and real-time data
Archiving of unstructured data	Well files, field development reports, drilling records, core Images, external studies, completion, and workover reports

with time. This integration requirement means that the data infrastructure and architecture must be defined and configured before rolling out the DOF program, supported with strong information management and indexing of data.

Table 4.2 shows a decision-support framework for determining the business value for data integration. With the assumption that we begin from nonintegrated data, interpret the table as follows:

- No data integration is needed if the amount of data is <100 TB, associated Health, Safety, and Environment (HSE) risk is zero, investment in data management is less than $10 million, predicted uncertainty reduction via integration is less than two points, no associated professional service automation (PSA), the project has reached the end of license and portfolio size exceeds more than the 10 biggest assets.
- Otherwise the business decision should be taken to pursue data integration; however different combinations of decision attributes from Table 4.2 are also viable. For example, one possibility is to track the value

Table 4.2 Decision Support Framework for Business Value of Data Integration

	Quantity	HSE/Risk	Finance	Uncertainty Reduction	Contract	Project Timeline	Portfolio Size
No integration	0–100 TB	0% Change	<$10 Million	<2 Points	No PSA associated	End of license	>10 biggest assets
Integration	Above 100 TB	>1% Change	>$10 Million	>2 Points	PSA associated	Early stage	2 biggest assets

of data integration to examine when an authorization for expenditure (AFE) was created, and then perform history matching against previous costs and revenue to see if any added value has been created.

4.1.1 Artificial Intelligence: Overview of State of the Art in E&P

Artificial intelligence (AI) techniques have been used in the E&P industry since the early 1970s (Bravo et al., 2014). After several decades of R&D and focused implementation—through smart wells, intelligent fields, expert systems and real-time analysis, and interpretation of large-scale data for process optimization—AI is now maturing in E&P. A literature search indicates that there is no unique consensus on AI techniques commonly used in the E&P community; however, artificial neural networks (ANNs), fuzzy-cluster analysis, evolutionary (genetic) algorithms, genetic optimization, and fuzzy inference analysis appear to have had a predominant role in applications in reservoir modeling and simulation, production and drilling optimization, drilling automation, and process control (Braswell, 2013). For example, Mohaghegh (2005) combines most of the aforementioned AI techniques under integrated intelligent systems, dividing them into four main categories: fully data driven (e.g., developing synthetic well logs), fully rule based (e.g., well-log interpretation), optimization (e.g., history matching), and data/knowledge fusion (e.g., candidate-well selection).

However, with the recent expansion of intensively harvesting the hyper-dimensional, complex, fast/streaming, and Big Data from oil and gas assets, AI techniques are increasingly seen as compatible with the methods of predictive data analytics. Recently, the term *artificial intelligence and predictive analytics* (AIPA) was coined (Bravo et al., 2014), which puts AI techniques into a broader context of techniques for data and business analytics, data mining, process control, automation and optimization, and advanced visualization. Bravo et al. (2014) provide a comprehensive summary of AIPA families and techniques, captured in Table 4.3. Selected AIPA techniques are described later in this chapter.

While the E&P industry is systematically heading toward comprehensive model integration between static, dynamic, surface and the entire production systems, AI is already being deployed to identify model inconsistencies, narrow model, and process uncertainties; improve forecasts and option assessment; mitigate risks; and support better decision-making. Moreover, with the worldwide implementation of DOF programs, the application of AIPA techniques is also increasing.

Table 4.3 Summary of AIPA Families and Techniques.

Family	Specific Technique
Computational intelligence	Neural networks
	Fuzzy systems
	Evolutionary computation
Data mining	
Rule-based case reasoning	Bayesian networks
	Expert systems
Automatic process control	Classical
	Robust
	Adaptive
	Intelligent
	Stochastic
Workflow automation	
Proxy models	Surrogate models
	Top-down models
Virtual environments	

From Bravo, C., Saputelli, L., Rivas, F., Pérez, A.G., Nikolaou, M., Zangl, G., et al., 2014. State of the Art of Artificial Intelligence and Predictive Analytics in the E&P Industry: A Technology Survey. SPE 150314-PA, https://doi.org/10.2118/150314-PA.

In 2009, the Society of Petroleum Engineers (SPE) (the E&P flagship professional organization) have established the AIPA subcommittee, within its Digital Energy Technical Section, with the mission of promoting the development and application of AIPA techniques in the oil and gas industry. With increasing interest and uptake of AIPA technologies in oil and gas, in 2011, the subcommittee was promoted to a new technical section, named Petroleum Data-Driven Analytics (PD2A). Bravo et al. (2014) have conducted a comprehensive technology survey that provides the state of the art of AIPA use in the oil and gas industry.

According to approximately 75% of respondents, management of large volumes of data remains a major challenge of the E&P industry, mostly because of the lack of integration in IT management and analysis. While automated process control is perceived as the most productive and mature AIPA technology in DOF programs worldwide, Fig. 4.1 indicates that data mining, neural networks, workflow automation, fuzzy logic, and expert systems are the most recognized AIPA applications.

In particular, data mining appears to be the most familiar AIPA technology, mostly in areas of data management and integration, data filtering, cleansing and imputation, and information search.

Data mining	65
Neural networks	58
Workflow automation	47
Fuzzy logic	45
Expert systems	42
Automatic process control	40
Genetic algorithms	36
Rule-based on reasoning	34
Proxy models	31
Virtual models	31
Machine learning	21
Intelligent agents	19
None	10
Others	4

Fig. 4.1 Professional awareness of AIPA technologies in oil and gas industry. Numbers are given in percent (%). *(Modified from Bravo, C., Saputelli, L., Rivas, F., Pérez, A.G., Nikolaou, M., Zangl, G., et al., 2014. State of the Art of Artificial Intelligence and Predictive Analytics in the E&P Industry: A Technology Survey. SPE 150314-PA, https://doi.org/10.2118/150314-PA.)*

On the other hand, statistical and machine learning (ML) techniques [which is one of the fastest growing technical fields and in the core of AI and evidence-based decision-making data science in health care, manufacturing, education, financial modeling, policing, marketing, and even social networking (Jordan and Mitchell, 2015)] remain relatively underutilized in E&P. The results of the survey suggest that the reasons for this underutilization may be attributed mostly to the relative obscureness and advanced technical concepts of ML, with the limited sources of information available for engineers and geoscientists; however, the situation is improving.

Lochmann and Brown (2016) further argue that the concepts of "intelligent energy," which largely encompass the methods and techniques of AIPA, have reached a strategic inflection point (SIP) in the oil and gas industry as "numerous case studies have documented new ways of working and more-than 10-folds improvement to individual productivity, demonstrating that new, more-effective ways of operating oil and gas assets are possible and practical."

4.1.2 Data Analytics: Descriptive, Diagnostic, Predictive, Prescriptive, and Cognitive

In today's data-intensive world, four types of analytics are available to help companies better harness the value of information (VOI) and knowledge hidden in data. Gartner provides an excellent example of classification of different types of analytics from the perspective of added value and related complexity of implementation and use, where Laney (2012) distinguishes between four main concepts of analytics (Fig. 4.2): descriptive, diagnostic, predictive, and prescriptive.

- *Descriptive analytics* is an approach to help us understand and answer the question "what happened?" during a given past period and verify whether or not a campaign or an action was successful. It is based on simple parameters such as number of trials or repetitions of a certain job. >90% of companies today (oil and gas and other industries) use this type of very basic analytics. However, even when using simpler, descriptive analytics, the data must be explored, visualized, comprehended, and

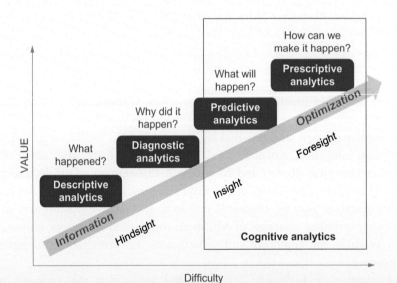

Fig. 4.2 Four different types of data analytics from the perspective of added value vs. related complexity of implementation. The red box was added to capture predictive and prescriptive types of data analytics that are frequently associated with the so-called *cognitive analytics*. *(Modified from Laney, D., 2012. Information, Economics, Big Data and the Art of the Possible With Analytics. Presentation by Gartner Inc., https://www-01.ibm.com/events/wwe/grp/grp037.nsf/vLookupPDFs/Gartner_Doug-%20Analytics/$file/Gartner_Doug-%20Analytics.pdf.)*

interpreted in the decision-making process. Analyzing large data sets is inefficient and useless without powerful visualization tools and techniques (explained further in Section 4.2.3). Hence, descriptive analytics leverages heavily the concepts of exploratory data analysis (EDA) (Gelman and Hill, 2007; Seltman, 2015), which integrates advanced and interactive charting and graphing with concepts of univariate statistics (e.g., plotting statistical distributions and modes, such as histograms, distributions, mean, variance, confidence intervals, etc.), bivariate statistics (e.g., cross plots, Q-Q plots, box plots, R^2, covariance statistics, and spatial variogram-based analysis) and multivariate statistics or analysis (MVA) (Tabachnick and Fidell, 2013), which combines analytical and visualization techniques such as principal component analysis (PCA), factor analysis, multidimensional scaling (MDS), or data clustering.

- *Diagnostic analytics* helps us determine the "root cause" of certain outcomes. While traditional key performance indicators (KPI) can provide a quantitative measure of performance, getting additional insight into "why something happened" requires diagnostic business intelligence tools. However, diagnostic analytics is laborious and frequently burdened with hindsight bias (choosing data that matches results); it provides an improved understanding of a limited piece of the problem we want to solve. For example, we can build an analytics dashboard for providing information on the root-cause analysis of electric submersible pump (ESP) failure events. The dashboard could show the basic linear causal relationships among variables; however, it would fall short in capturing complex nonlinear variable correlations. Studies show that <10% of companies surveyed do this type of analysis on occasion, and <5% do so consistently.

The next two classes or types of data analytics are usually the ones referred to by the technical and business analytics experts as the analytics types that can really provide the insight and foresight into how to drive the technical and business decisions forward. Predictive and prescriptive analytics are considered branches of the so-called cognitive analytics, because they combine elements, methods, and tools of cognitive science, such as AI, statistical inference, ML, and multimodal deep learning (DL) of visual, language, and knowledge recommendations.

- *Predictive analytics* provides the ability to use data (structured and unstructured) to derive patterns and forecast future events and outcomes with mathematical certainty. It helps us discover hidden, nonintuitive patterns in Big Data and understand complex causal relationships and correlations, mostly of a highly nonlinear nature (see Table 4.3 for a classification of the main tools and techniques gathered under the umbrella of AIPA). Examples of predictive analytic implementations include modeling, equipment operations, production trends, reservoir dynamics, and asset failure predictions to minimize downtime. Financial predictive algorithms compute expected response and ROI for workovers, pump changes, injection rate modifications, drilling plans, and alternative completion parameters. However, recent surveys indicate that less than 1% of companies surveyed have actually deployed predictive analytics. The ones who have found extremely encouraging results that have added significant values to their businesses.

- *Prescriptive analytics* apply the outcomes and insights of predictive analytics and turn them into actionable foresights by applying advanced process optimization methods. Accurate predictions help us understand the actions to be taken to maximize good outcomes and minimize or prevent potentially bad outcomes. Examples of prescriptive analytics include alerts (e.g., opportunities, abnormalities, and data problems), recommendations (e.g., the workover to perform next, when to stimulate a well, optimal injection rates throughout a waterflood, or where to drill), and optimization (e.g., capital allocation, investment, and risk management). Currently, R&D in predictive analytics for E&P is cutting edge; for example, a recent breakthrough includes a novel data—physics paradigm in modeling and optimization of oil and gas assets (Sarma and Leport, 2016).

To conclude this section, we propose a schematic example of a modular advanced data analytics workflow for the E&P industry, where components of individual analytics domains, from descriptive to prescriptive, merge into a collaborative synergy (Fig. 4.3). The workflow consists of five modules and begins with the database management module, which includes data acquisition (i.e., data from smart sensors/IoT, etc.), integration [i.e., subsurface (geological and geophysical), drilling and completions, production, stimulation, operations], and aggregation for the purposes of statistical analysis.

Module 2 combines data exploration steps that include EDA, with uni-, bi-, and multivariate statistical analysis, and examination of the most important variables for the predictive model. Module 2 also combines missing data and outlier analysis and temporal/spatial smoothing, which is often overlooked in the data preparation phase. Techniques like data imputation [e.g., multivariate imputation by chain equations (MICE) (van Buuren and Groothuis-Oudshoorn, 2011)] or interpolation [e.g., regression methods (such as LOESS/LOWESS), kriging, etc.] can be considered.

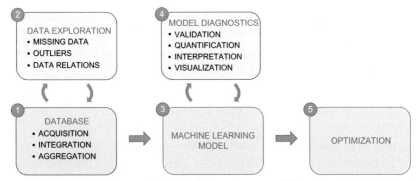

Fig. 4.3 Proposed modular advanced data analytics workflow for the E&P industry, where components of individual analytics domains, from descriptive to prescriptive, merge into a collaborative synergy.

The workflow continues with Modules 3 and 4, which combine selection and building the predictive analytics (e.g., ML) model (for ML model selection, see Section 4.2) as well as validation, quantification, interpretation, and visualization of results. The workflow ends with Module 5 and the prescriptive analytics phase, where the results of the predictive model provide an input for a nonlinear optimization, where certain KPIs can be defined as minimization [e.g., cycle time or nonproductive time (NPT)] or maximization (e.g., production, rate of penetration (ROP)) problems via a suitable objective/cost function [e.g., in sparse equations and least square (LSQR) form]. The benefits of deploying advanced data analytics workflow in the modular form are as follows:

- A project can grow in functionality by adding project files for tasks.
- Intellectual property (IP) can be modularized within individual project files, which makes collaboration easier.
- Modularizing promotes functionality reuse, unit tests, easier documentation, etc.

4.1.3 Big Data in E&P: Concepts and Platforms

E&P operations have traditionally generated large volumes of data; however, with the advent of "smart operations" and DOF projects, the E&P industry is now producing extreme volumes, at exponentially higher rates than ever before. Today's operations generate terabytes and petabytes of data, at extremely large volumes, speeds, and acquisition frequencies from multiple sources and domains, such as geophysical, geological, engineering, production, surveillance, maintenance, etc. The E&P industry is quite literally experiencing data and information overload; it needs a focus and

Fig. 4.4 The E&P industry needs processes to transform vast amounts of data into good operational decisions.

dedicated approach to transform these vast volumes of data into decisions, to devise processes to transform information into knowledge (Fig. 4.4).

However, according to recent studies and reports in Alain Charles (2015), the oil and gas industry has been using only 1% of the data it generates; which means that 99% of acquired data remains to be exploited to generate business value. With the evolution of the Big Data paradigm, E&P companies are focusing on mining value from this data, with the main objective of "getting value from all data by leveraging emerging technologies and pattern-based techniques for innovation, strategy, faster and better decisions" (Davis, 2015). Moreover, we increasingly hear the E&P industry saying that Big Data is now the new oil.

The gradual shift of the E&P industry to digital data-driven technology in oil fields is expected to improve the productivity of pipeline operation and safety by 30% (Alain Charles, 2015). Another striking example for potential improvements is in the pump performance. If globally, the industry improved pump performance and efficiency by even 1%, it could increase oil production by half a million barrels per day and generate an additional $19 billion of revenue per year. Moreover, the global oil and gas industry is facing major challenges where improved data analytics could help, for example, extraction costs are rising and the market has been affected by a dramatic drop in oil prices and the turbulent state of international politics, which adds to the uncertainty in exploration and drilling for new reserves. To help address these and other challenges, key companies in E&P are looking to Big Data in search of maximum optimization at minimum cost. Fig. 4.5 shows the main areas of interaction between attributes of Big Data analytics and E&P business segments with the most potential to add value.

In the current literature, experts usually refer to Big Data in terms of "the Vs." Brulé (2013) and Davies (2014) categorize Big Data in terms of 3 Vs—volume, velocity, and variety. The Oil Review (2015) considers 4 Vs, by adding *value*, and Davis (2015) goes a few steps further and defines 7 Vs as the 7 pillars of Big Data, which include the following:

• *Volume*: this component of Big Data addresses massive quantities of acquired data, rising from terabytes to petabytes. Traditionally, the data are collected and loaded into data warehouses. With Big Data, the focus

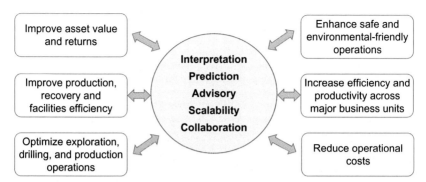

Fig. 4.5 Main areas of interaction between the attributes of Big Data analytics and E&P business segments, with the most potential to add value.

is to extract, load, and transform. The new paradigm is to collect and load data into the Apache Hadoop open source database (Ghemawat et al., 2003; Handy, 2015), which enables distributed processing of large data sets on clusters and servers, without extensive transformation into a relational database model for further analysis.

- *Velocity*: this component relates to the understanding that the acquired data are no longer data at rest (or static) and adopting new methods for data in motion (e.g., streaming or fast data) to analyze data in real time. Not all data received in real time need real-time analysis. However, some (e.g., real-time alerts for operational efficiency and failure diagnostics) need real-time adaptive analytics with stream computing and support of massively parallel-processing databases (Brulé, 2009) and low-latency data-flow architecture (Brulé, 2013).
- *Variety*: Big Data consist of structured and unstructured data. While structured data are generally in digital form, acquired by sensors (e.g., temperature, pressure, fluid flow), the unstructured (no format) data come in the form of text files, well files, field development reports, drilling records, etc. (see Table 4.1) and requires specific types of text analytics (down into Boolean operands) to extract information at large scales.
- *Veracity*: this component relates to the accurateness and correctness of data. In circumstances of the first 3 Vs, confusion can arise because of incomplete (and sometimes obscure) definitions of how true and trustworthy are the data?
- *Virtual (data)*: this component of Big Data enables the E&P industry to generate abstracted and integrated information in real time, from disparate sources, and send it to multiple applications and users. The virtual data centers/servers are easier to build and consume (than traditional data stores), and require much less effort to maintain.

- *Variability*: this Big Data attribute can occur in each of the other six pillars. The range of variability (not to be confused with the more widely adopted uncertainty) depends on the stage of an E&P program under consideration as a function of time, location, or some other measured parameter. Moreover, Begg et al. (2014) refer uncertainty to not knowing the value (or answer) of some quantity and define variability as the multiple values the quantity has at different locations, times or instances.
- *Value*: this attribute clearly represents the most important component of Big Data, measured in monetary or nonmonetary capacity.

In the transcript of a recent survey by Accenture and Microsoft, focusing on 2016 E&P digital trends, Holsman and Richards (2016) have reported that almost 90% of responding E&P companies, despite the industry downturn, plan to maintain or increase investments in digital technologies, predominantly in Big Data-powered analytics, the IoT, and cloud-enabled mobility. However, though the survey results indicate that more than half of respondent believe that digital technologies have added significant value to their businesses, the general impression is that Big Data analytics are still being widely underutilized in the oil and gas industry. Although identified as a key capability that E&P needs to leverage, only 13% of survey respondents felt that their companies had mature analytics capabilities, and almost two-thirds said that their companies would be investing in analytics over the next years to close the gap.

Traditionally, the E&P industry-standard approach for data analysis has been heavily leveraging mainstream spreadsheet-based tools and basic, macro- or script-enabled workflow automation. With the new paradigm of Big Data and the associated complexity as outlined previously in the description of the 7 Vs, the traditional data analysis tools and techniques quickly become suboptimal, due to intricate, nonlinear, multivariate, and nonintuitive root-cause data relationships that affect decision-making.

At SPE Forum series event, "Next Generation of Smart Reservoir Management: The Eminent Role of Big Data Analytics," held in 2016 in Dubai, UAE, the authors have concluded that sustainable transformation of E&P businesses to fully harness the potential of Big Data requires scalable data analytics with cognitive abilities, like massively distributed data mining and machine and statistical learning, with little or no human supervision. This is briefly addressed in the following sections. An efficient implementation of Big Data analytics can only be enabled by the innovative IT and data management solutions that allow access to all data, all the time, and by all stakeholders. This type of access increasingly being provided through the implementation of flexible, open data management architectures such as

Apache Hadoop (Ghemawat et al., 2003; Handy, 2015) or NoSQL (Pokorny, 2011), distributed on platforms such as Cloudera, Hortonworks and MapReduce (Dean and Ghemawat, 2008) or Apache Spark.

Recently, an overwhelming amount of literature has been published about Big Data concepts. Two publications that we recommend include "Harness the Power of Big Data" by Zikopoulos et al. (2013) and *Harness Oil and Gas Big Data with Analytics: Optimize Exploration and Production with Data Driven Models* by Holdaway (2014).

4.2 INTELLIGENT DATA ANALYTICS AND VISUALIZATION

4.2.1 Data Mining

Data mining (DM) is a knowledge discovery from large quantities of data. The process derives its name from the similarity between searching for valuable business information in a large database, containing terabytes or even petabytes of data, and mining a mountain for a vein of valuable ore. Technically, the term refers to the process of extracting useful models and patterns that are (Leskovec et al., 2014)

- valid (i.e., contain new data with some certainty),
- useful (i.e., add value and enable people to take related actions),
- unexpected (i.e., nonobvious and nonintuitive, spurring the "aha!" moment), and
- understandable (i.e., humans should be able to interpret and analyze them).

Data mining as a discipline overlaps with database systems, statistics, and ML, and, as such, the complexity when dealing with data in data mining applications can be graphically represented as shown in Fig. 4.6.

As data come in a variety of modalities, formats, and ontologies—from structured, unstructured, static to streaming, descriptive to Boolean—this infers that for successful data mining, the data need to be properly collected, stored, and managed. Ideally, these tasks would be performed continuously by the data operators; however, in reality (as is frequently the case in the E&P industry), the data presented for mining is imperfect, with missing, illogical, and nonphysical values that require extensive QA/QC processing, with missing data interpolation and imputation (van Buuren and Groothuis-Oudshoorn, 2011).

Historically, statisticians were the first to use the term "data mining," ironically focusing primarily on the attempts to extract the information that was not supported by the data. However, with the evolution of statistical

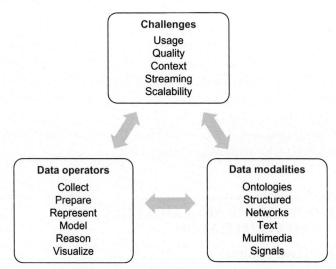

Fig. 4.6 Importance and relevant areas that interact in data mining projects. *(Modified from Leskovec, J., Rajaraman, A., Ullman, J.D., 2014. Mining of Massive Datasets, second ed., Cambridge University Press.)*

modeling and high-performance computing (HPC), modern data mining is largely extracting data models or patterns that can sometimes be the summary of the data or even the set of most extreme features of the data. The data mining tasks are mainly classified into the following:

- *Descriptive methods*: where automated and intelligent tools discover previously unknown human-interpretable patterns that describe the data.
  - Example: data clustering of reservoir parameters to identify sweet spots for new drilling campaigns (Roth et al., 2013).
- *Predictive methods*: where automated systems [e.g., recommendation system (Leskovec et al., 2014; Jordan and Mitchell, 2015)] and models use certain variables (predictors) to predict unknown future values, trends, or behavior of other (response) variables.
  - Example: well production prediction and optimization (Zhong et al., 2015) or equipment predictive maintenance, both based on historically recorded data.
- *Root-cause analysis*: where automated tools are used to identify roots and causes of a system's faults and problems, mostly based on the analysis of historical categorical, continuous, and temporal data.
  - Example: down-time or job-paused time analysis of hydraulic fracturing, well artificial lift, or well stimulation equipment (Maucec et al., 2015).

Although the data mining is rapidly gaining rightful popularity—particularly in conjunction with Big Data analytics, where DOF applications are certainly not an exception—a caveat related to the risk that a data mining analyst may discover patterns that are meaningless, because they are not supported by the data exists. Consequently, this effect, the statisticians call Bonferroni's Principle (Leskovec et al., 2014), may, for example, generate statistical artifacts rather than evidence of the conducted search and lead to unrealistic predictive models. The solution comes in the form of the Bonferroni correction, when several dependent or independent statistical tests are being performed simultaneously on a single data set.

4.2.2 Statistical and Machine Learning

Although the terms statistical learning and ML differ by name, they are quite similar, and, in fact, both types of learning are inseparably intertwined. Statistical learning refers to the set of tools for modeling and understanding complex and large-scale data sets, such as Big Data. It is a fairly recently developed area of statistics and largely complements the developments in computer sciences (e.g., advanced data management and cloud computing) and ML. ML addresses the question of "how to build computers that improve automatically through experience" (Jordan and Mitchell, 2015). This section gives a brief overview of the core ML methods and outlines some trends and prospects for future developments. It summarizes the most popular ML techniques, highlights its three main paradigms, and provides characteristic examples. As ML is becoming increasingly popular in the E&P industry, a few successful applications relevant to the DOF are presented in Section 4.3.

Conceptually, ML algorithms can be viewed as navigating through a large domain of candidate programs to identify a program that optimizes a specified performance metric or objective. The application of ML algorithms varies greatly depending on the nature of the problem, for example, through use of decision trees, mathematical functions, optimization, etc. However, with the vast amount of Big Data, it is imperative that the common denominator of ML techniques appropriate for DOF applications become highly scalable solutions which support the platforms of the cloud and HPC, real-time analytics, and the rapidly expanding IoT, all with robust and resilient cybersecurity mechanisms (see Chapter 2, *Instrumentation and Measurement*). For more information see *The Elements of Statistical Learning: Data Mining, Inference and Prediction* by Hastie et al. (2011), *An Introduction to Statistical learning: with applications in R* by

James et al. (2014), and *Jump-start Machine Learning in R: Apply Machine Learning with R Now* by Brownlee (2014). The latter two references focus specifically on applications of the ML algorithms and techniques in programming language R, which has become a de facto standard for statistical computing (The R Foundation, 2017).

ML methods can be classified into three main paradigms: supervised learning, unsupervised learning, and reinforcement learning (RL) (Jordan and Mitchell, 2015), which are summarized below and in Table 4.4.

- *Supervised learning.* Let us assume an ML system with a set of input parameters $(x_i; i = 1,...,n)$, called *predictors* and associated output/measured variables called the *responses* (y_i). The supervised learning system generally yields its prediction via a learned mapping function $f(x)$, which produces an output y_i for each x_i or a probability distribution $p(y|x)$. The objective is to design and fit a model that finds a relation between the response and predictors, with the objective of accurately predicting the response for future observations (predictions or forecasts). In supervised learning, response variables are usually characterized as *quantitative* (also referred to as *continuous*) or qualitative (also known as *categorical*). It is a common practice in data science to refer to problems with a quantitative response as *regression problems* and to those with a categorical response as *classification problems*. A variant of regression and classification methods are the so-called *tree-based methods*. These methods work on the principle of segmenting the predictor space into a number of simple regions. To make a prediction for a given observation, the tree-based methods usually use statistical moments such as mean or variance. As the set of splitting rules to segment the predictor space can be conveniently represented as a tree, it visually makes the decision process significantly easier; these types of approaches are also referred to as *decision trees methods*.
- *Unsupervised learning* addresses more challenging situations, where for every observation $i = 1,...,n$, one finds a vector of measurements x_i but no associated response y_i. Hence, it is not possible to fit a linear regression model because there is no response variable to predict. In such conditions, finding a solution is less transparent and the approach is referred to as *unsupervised*. The two main classes of unsupervised learning methods are the *cluster analysis* or *clustering* and the so-called *dimensionality reduction* methods. The objective of cluster analysis is to determine, on the basis of variables or parameters $x_1, ..., x_n$, whether these observations can be classified into relatively distinct groups called *clusters* and if there is a possibility to represent individual clusters with their single

Table 4.4 Summary of the Most Popular Methods and Classifiers in ML, as
Representative of the Three Main Paradigms Described Previously, With Some
Suggested References for Further Reading

	ML Family		
ML Paradigm	**Type of Problem**	**Technique**	**Suggested Reading**
Supervised learning	Regression	Linear regression (LR) • Ordinary least squares • Stepwise and moving LR Penalized LR • Ridge LR • Elastic nets Nonlinear regression • Multivariate adaptive regression splines (MARS) • Support vector machine (SVM) • K-nearest neighbor • Neural network (NN) Decision trees for regression • Classification and regression trees (CART) • Conditional decision trees • Bagging CART • Random forest (RF) • Gradient boosted machine (GBM)	Hastie et al. (2011) James et al. (2014) Leskovec et al. (2014) Brownlee (2014)
	Classification	Linear classification • Logistic regression • Discriminant analysis Nonlinear classification • Mixture, regularized, quadratic and flexible discriminant analysis • Support vector machine (SVM) • K-nearest neighbor • Naive Bayes Nonlinear classification with decision trees • Classification and regression trees (CART) • Bootstrapped aggregation (Bagging) CART • Random forest (RF) • Gradient boosted machine (GBM)	

Continued

Table 4.4 Summary of the Most Popular Methods and Classifiers in ML, as Representative of the Three Main Paradigms Described Previously, With Some Suggested References for Further Reading—cont'd

ML Paradigm	Type of Problem	Technique	Suggested Reading
	ML Family		
Unsupervised learning	Clustering	• K-means • Hierarchical	James et al. (2014) Hastie et al. (2011) Leskovec et al. (2014)
	Dimension reduction	• Principal component analysis (PCA) • Factor analysis • Multidimensional scaling (MDS)	Tabachnick and Fidell (2013)
Reinforcement learning		• Markov decision process (MDP)	Sutton and Barto (1998)

representative elements, called *centroids*. With dimensionality reduction methods on the other hand, the analyst is aiming to represent the complex numerical model with a reduced or compressed set of (principal) components (usually referred to as *eigenvalues* and *eigenvectors*), which can still adequately represent the observation domain $x_1, \ldots, x_n$, while significantly reducing the computational effort and complexity. The clustering and dimensionality reduction methods often fall under the category of MVA methods (Tabachnick and Fidell, 2013).

• *RL* is a paradigm where the information available in the training of the ML model can be viewed as a cross-section between supervised and unsupervised learning. The RL methods usually leverage the ideas and algorithms from the area of control theory (e.g., optimal, robust control, closed-loop control, and variance reduction). The mathematical foundation of ML methods represents Markov decision processes (MDP), similar to Markov chains and its Monte Carlo approximations. The complexity of these techniques goes beyond the scope of this book; to learn more, see Sutton and Barto (1998).

It is interesting to note that in the applications of advanced data analytics, a small number of methods always seem to perform better than most others. This phenomenon has been documented for classification problems (Fernandez-Delgado et al., 2014) in a study that combined 179 classifiers from 17 families, while using 121 data sets. In petroleum engineering and geoscience applications that largely pertain to DOF architecture as well, many of Big Data problems involve regression. However, it seems like the same set of methods always outperform others: (artificial) neural network, support vector machine (SVM), and random forest. Depending on the nature of the application, some of these methods are more suitable than others, but only rarely does the user need to look outside of this suite to more exotic methods. Section 4.3 briefly captures some of the prominent applications of data analytics from the E&P domain, pertaining to DOF projects.

Although the field of ML is a relatively young one, it is rapidly growing. One of the emerging trends in ML is the so-called recommendation systems (Leskovec et al., 2014; Jordan and Mitchell, 2015), applications that involve predicting user responses to different options. One genre of recommendation systems is the area of *collaborative filtering*, which recommends items or actions based on similarity measures between the users and/or items. Such a paradigm naturally appeals to the concept of DOF, which is by definition a collaborative environment [e.g., real-time operation center (RTOC)], where oilfield operators work with advanced sensor- and data-driven technology.

Maybe the future of ML techniques for DOF and data-driven E&P operations is hidden in the analogy to natural learning systems. As envisioned by Jordan and Mitchell (2015), this concept suggests the idea of team-based, mixed-initiative learning. In a nutshell, since the current ML systems mostly operate in isolation to analyze given data, people often work in teams to collect and analyze data, by bringing together a variety of expertise and perspectives when solving particularly complex and difficult problems (e.g., large-scale integrated reservoir studies). Perhaps in the next-generation DOF, new ML methods will work collaboratively with oil and gas field operators to extract and mine deep knowledge and subtle statistical regularities from massive data sets (Big Data) acquired at extreme velocities and frequencies by smart IoT sensors, to generate intelligent real-time operational decisions.

For completeness and to facilitate easier understanding of the applications presented in Section 4.3, we present a quick overview of the mentioned ML techniques.

4.2.2.1 Artificial Neural Network

ANNs are ML systems based on the workings of the brain, which is known to consist of a massively interconnected system of neurons that do sensory processing, control motor functions, and engage in patterns of thought. ANNs are "trained" by a large number of input patterns that cause them to "learn" from the experience from the bottom-up approach. The structure of a neural network is usually drawn as a hierarchy of layers (input, hidden, and output) in which nodes (representing neurons) are connected by arcs (see Fig. 4.8A). The arithmetic value of any node is equal to the sum of the values of the preceding nodes each multiplied by the weight of the connecting arc, called the *activation function*:

$$y_i = \Sigma w_{ij} x_j \tag{4.1}$$

where y_i is the value of the ith node, x_i the value of the jth node of the preceding layer, and w_{ij} is the weight associated with the arch that connects the two nodes. The output node is governed by the activation function and a threshold that determines the initiation of output. In simpler networks, a node fires and passes output when the node value, y_i, exceeds a given threshold value, U. The firing state of a node is either 1 or 0, determined by whether the activation, a, is positive or negative, where (Fig 4.7A)

$$a = y_i - U \tag{4.2}$$

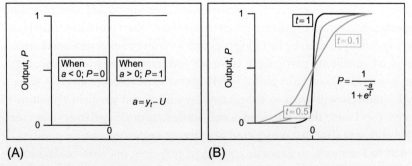

(A)   (B)

Fig. 4.7 Control of the firing of a neural network node by an activation function using: (A) threshold value determination of positive activation and (B) sigmoidal function, where the value of t influences the relative rate of activation. *(Modified from Doveton, J.H., 1994. Geologic Log Analysis Using Computer Methods. AAPG Special Volumes.)*

More recent networks commonly use a sigmoidal function to model the transfer between input and output signals (Fig 4.7B):

$$P = \frac{1}{1 + e^{\frac{-a}{t}}} \qquad (4.3)$$

where P is the probability of the node firing, t a constant that determines the function steepness, and a is the activation node. The steepness of the activation function determines whether most of the input is transferred through the nodes, or whether the output is only initiated by stronger inputs. This feature attempts to imitate the behavior of real neurons, which often tend to be either active or inactive.

A basic model often uses three layers of nodes (Fig 4.8A): the input layer receives the data, the middle (hidden) layer draws stimulation from the input layer, and transmits onward to the final output layer, which is the result of the system. In "training" the network, a set of patterns is repeatedly presented and the weights of the arcs are modified such that the output makes a better match with a desired result. The "training" is usually accomplished by backward propagation of errors through the network that distributes the difference between the desired result and the actual output as small incremental adjustments in the interconnection weights. The process is gradual and iterative, until the weights converge to an equilibrium setting and the network is trained. The speed of the training is controlled by a learning rate set by the user. If too high, the network learns quickly, but the weights may oscillate with an unstable solution. Very slow learning rates ensure a smoother passage to stability but may take excessive computing time.

At the end of the training, an unknown pattern can be entered for purposes of *classification* or *regression* (prediction) as outlined in previous sections. Fig 4.8B presents an example of an input–output model for application of ANNs developed for the optimization of hydraulic fracturing in unconventional gas reservoirs (Temizel et al., 2015).

4.2.2.2 Support Vector Machine

The SVM is widely perceived as one of the most powerful classification "out-of-the-box" learning tools. It has been developed in the area of computer science in the early 1990s and has recently been receiving more and more attention in the widest range of engineering fields, mostly because of the advent of Big Data applications. According to James et al. (2014), the SVM is a generalization of a simple and intuitive classifier called a maximum margin classifier, which unfortunately cannot be applied to most data

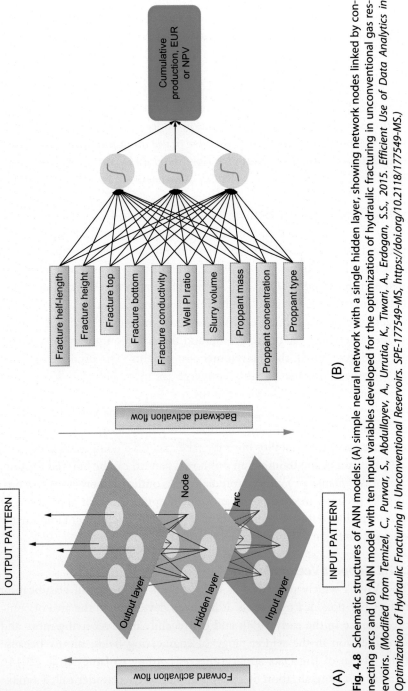

Fig. 4.8 Schematic structures of ANN models: (A) simple neural network with a single hidden layer, showing network nodes linked by connecting arcs and (B) ANN model with ten input variables developed for the optimization of hydraulic fracturing in unconventional gas reservoirs. *(Modified from Temizel, C., Purwar, S., Abdullayev, A., Urrutia, K., Tiwari, A., Erdogan, S.S., 2015. Efficient Use of Data Analytics in Optimization of Hydraulic Fracturing in Unconventional Reservoirs. SPE-177549-MS, https://doi.org/10.2118/177549-MS.)*

sets because it requires the classes to be separable by a linear boundary. The support vector classifier is an extension of a maximum margin classifier that can be applied to a broader range of problems. Finally, the SVM is a further extension and next-generation support vector classifier and can accommodate nonlinear class boundaries. These three terms are often loosely and interchangeably used, but it is important to distinguish between them when deploying the SVM method.

The mathematical details behind the derivation of both the maximum margin classifier and the support vector classifier are beyond the scope of this book; for more information see James et al. (2014) and Leskovec et al. (2014). However, the main difference between the support vector classifier and the SVM, as a matter of our interest, is shown in Fig. 4.9. Let us assume

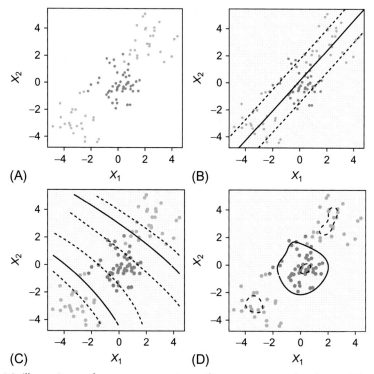

Fig. 4.9 Illustrative performance comparison of support vector classifier and the SVM on a nonlinear classification: (A) observation data arranged in two classes, colored in red and blue, (B) relatively poor classification performance of support vector classifier due to defined linear boundaries, (C) significantly better fitting classification using SVM with third-order polynomial kernel, and (D) superior fitting classification using SVM with the radial basis kernel. *(With permission from James, G., Witten, D., Hastie, T., Tibshirani, R., 2014. An Introduction to Statistical Learning with Applications in R. Springer, NY.)*

we are solving a classification problem where the observation data fall into two classes, colored in pink and blue in Fig. 4.9A. The support vector classifier seeks the linear boundary between the two classes of observed data points and thus performs quite poorly (Fig. 4.9B).

When applying the SVM method, which is an extension of support vector classifier, the classification feature space is rearranged in a specific way using nonlinear functions, that is, kernels. If then, an SVM with a polynomial kernel of the third degree is applied to the nonlinear distribution of data points as shown in Fig. 4.9A; the result is the significantly better fitting classification presented in Fig. 4.9C, which renders better decisions. Furthermore, if instead, the SVM is applied with the radial basis kernel, the classification/decision boundary is captured even more accurately (Fig. 4.9D).

When applied to statistical regression problems, the SVM method is referred to as support vector regression (SVR). Both techniques are closely related and only applied to a different class of problems. In the case of SVR, the regression function usually has the form (Zhong et al., 2015)

$$f(\mathbf{x}) = \gamma = \sum_{i=1}^{N} (\alpha_i^* - \alpha_i)(v_i^t x + 1)^p + b \qquad (4.4)$$

where $v_1, \ldots, v_N$ are N support vectors and b, p, α_i, and α_i^* are the parameters of the model, which are optimized with respect to ε-insensitive loss (Zhong et al., 2015). During the parameter estimation, the N support vectors are selected from the data training set. Similar to the nature of classification problems, solve the nonlinear regression with the application of kernel function. For information, the radial basis kernel function (as mentioned previously for classification) acquires the form:

$$K(v_i, x) = \exp\left(-\gamma |v_i - x|^2\right) \qquad (4.5)$$

4.2.2.3 Random Forest

Random forest (RF) is an ensemble ML method that constructs a large number of uncorrelated decision trees based on averaging random selection of predictor variables. For in-depth introduction into the concept of decision trees, see James et al. (2014). In their fundamental formulation, decision trees have proven to be very successful in solving classification problems of statistical learning; however, they are less efficient for nonlinear regression.

Various techniques have been introduced by statisticians to improve upon statistical learning capabilities of decision trees, like bootstrap aggregation or bagging (James et al., 2014); however, while bagging dramatically improves the prediction accuracy of decision trees, it comes at the expense of interpretability.

The RF approach provides an improvement over the bagged trees by de-correlating the trees, which reduces the variance when the trees are averaged. When building decision trees (they are generated in parallel), each time a split in the tree is considered and a random selection of m predictors is chosen as a subset of split candidates from the full set of predictors. Hence, as the new selection of m predictors is generated at each split, and one typically chooses $m \approx \sqrt{p}$, which means that the number of predictors considered at each split (m) is approximately equal to the square root of the total number of predictors, p.

The predictor variables for RF method can be of any type: numerical, categorical, continuous, or discrete. The method automatically includes interaction among the predictor variables in the model because of the hierarchical structure of trees. The fact that the RF trees are insensitive to skewed distributions (i.e., do not require mapping into normal score domains), outliers, and missing values (i.e., data imputation methods are less required), they are considered as one of the most efficient "of-the-shelf" predictive ML techniques.

4.2.3 Visualization and Interactivity

This section presents a summary and examples of selected diagrams, graphs, and images for qualitative and quantitative visualization of primarily predictive analytics, pertaining to statistical learning and ML and multivariate analysis (Fig. 4.10).

We did not have room to include examples for visualization of descriptive analytics (e.g., EDA with uni- and bivariate statistics, such as histograms, statistical distributions, box plots, QQ-plots, cross-plots, and correlation/covariance). For more information on these, see Gelman and Hill (2007) and Seltman (2015).

Fig. 4.11 shows example of interactive analytical dashboard visualization for DOF applications. The figure adds the Tornado chart (ranking importance of predictor variable in terms of response variable) and the receiver operating characteristics (ROC) (sensitivity of the binary classifier to false alarm probability) to the list of selected visualization options.

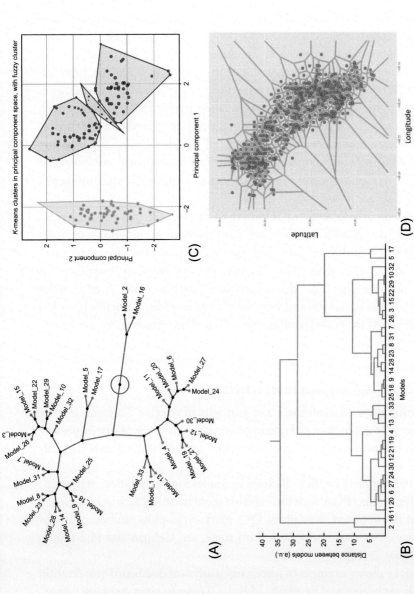

Fig. 4.10 (A) Constellation plot: hierarchical structure of simulation model realizations organized in four distinctive clusters; (B) hierarchical clustering: representation of simulation model hierarchy with dendrogram. The distance between model realizations is proportional to their dissimilarity in terms of model response. (C) K-means clustering: k-means clusters in principal component space, with fuzzy cluster in gray. (D) Voronoi diagram: spatial separation of the status of producing reservoirs. (E) Scatterplot matrix: selected predictor variables for horizontal well production optimization; and

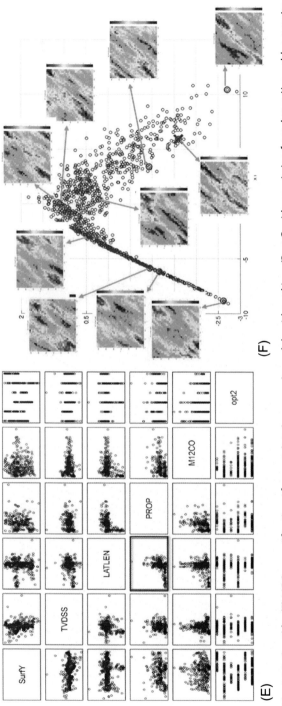

(F)

Fig. 4.10, Cont'd (F) MDS: quantification of uncertainty in reservoir models with ranking. (Part C: with permission from: https://www.blue-granite. com/blog/supply-chain-insights-with-advanced-analytics-fuzzy-clustering. Part D: with permission from: https://exchange.ai/downloads/teapot-dome-voronoi-chart/. Part E: with permission from Zhong, M., Schuetter, J., Mishra, S., LaFolette, R.F., 2015. Do Data Mining Methods Matter?: A Wolfcamp Shale Case Study. SPE-173334-MS, https://doi.org/10.2118/173334-MS. Part F: with permission from Caers, J., 2011. Modeling Uncertainty in the Earth Sciences. John Wiley & Sons, Ltd, Chichester, UK.)

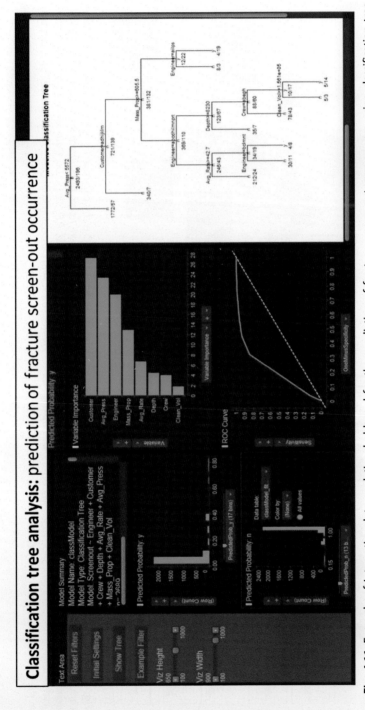

Fig. 4.11 Example of interactive data analytics dashboard for the prediction of fracture screen-out occurrences using classification tree analysis.

4.3 APPLICATIONS TO DIGITAL OIL AND GAS FIELDS

4.3.1 Machine Learning and Predictive Analytics

This section presents a few state-of-the-art and emerging applications of data mining and ML with predictive analytics in the E&P industry. The focus is on the applications of production optimization because they correlate more firmly with the data-driven reservoir management and are significant areas of application in DOF systems.

Mohaghegh et al. (2015) have introduced the concept of surrogate reservoir models (SRM) as a "smart" proxy for numerical reservoir simulation models. The SRM is an ensemble of multiple ML technologies, including pattern recognition and intelligent agents, which are trained to learn and consequently mimic the behavior of fluid flow physics using data generated by a numerical simulation model; however, SRMs run at extremely high speeds and complete the simulation run in a fraction of a second. Mohaghegh et al. (2015) have deployed the SRMs to increase field production and optimize choke size schedule—all without drilling new wells.

Bravo et al. (2014) have defined intelligent agents as computational systems comprising multiple active components that are capable of making decisions and taking actions autonomously. The intelligent agents are suited to processing large amounts of data in distributed environments and can also communicate and collaborate with each other to reach common objectives. For example, Zangl et al. (2011) have used intelligent agents in the form of self-learning expert systems to construct a holistic workflow for autonomous history matching. Fig. 4.12 (from Zangl et al., 2011) demonstrates a schematic of a hierarchical learning exercise of a history matching agent that, instead of accomplishing the change of state in sequential manner, splits a complex task (minimizing the Objective) into a set of isolated and repeatable subtasks executed at different layers of corrective actions.

In an SPE webinar, Saputelli (2015) has presented several varieties of supervised and unsupervised ML techniques (see Table 4.4) with optimization as emerging trends of predictive and prescriptive data analytics applications in E&P. However, ANNs are also being used in innovative applications. For example, Shirangi (2012) have built fast proxy models by combining ANNs and SVR models to solve a robust production optimization problem and used unsupervised ML (k-means clustering and MDS; see Table 4.4) to select an optimal set of representative reservoir model realizations. Recently, ANNs are being actively used in a variation known as a

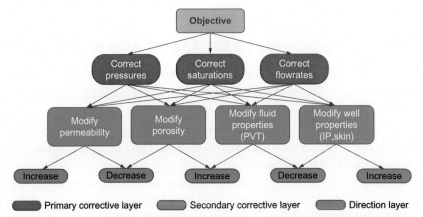

Fig. 4.12 Examples of a hierarchical learning exercise composed of intelligent agents to perform autonomous history matching. *(Modified from Zangl, G., Al-Kinani, A., Stundner, M., 2011. Holistic Workflow for Autonomous History Matching using Intelligent Agents: A Conceptual Approach. SPE-143842-MS, https://doi.org/10.2118/143842-MS.)*

Kohonen self-organizing map (SOM) (Fig. 4.13); it is referred to as a map because it assumes a topological structure among its cluster units and effectively maps cluster weights w_{ij1}, w_{ij2} ..., w_{ijn} to input data vector x_1, ..., x_n and generates output data vector y_1, ..., y_n.

Roy et al. (2013) are also using SOMs for supervised and unsupervised multi-attribute facies analysis in seismic stratigraphy, and Zangl and

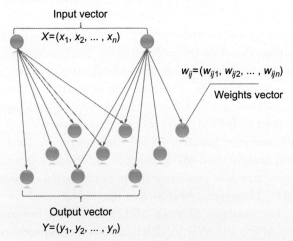

Fig. 4.13 A rendering of Kohonen self-organizing model architecture. *(Modified from Magomedov, B., 2006. Self-Organizing Feature Maps (Kohonen Maps), https://www. codeproject.com/Articles/16273/Self-Organizing-Feature-Maps-Kohonen-maps.)*

Stundner (2007) and Dossary et al. (2016) use SOMs to explore and identify regions in reservoir simulation models, based on geological signatures and/ or (dis)similarities. Such SOM models condition geology to reservoir flow dynamics and reduce simulation model computational requirements for inversion and assisted history matching (AHM) with minimal engineering effort. They propose an algorithm that stems from the original Kohonen algorithm but is redesigned to fit reservoir simulation data. As such, the algorithm is optimized for 3D spatial maps, rather than 2D data considered by the original Kohonen. The data "discovers" the underlying map rather than adapting to it, and the similarity metric is not the minimum distance Euclidean norm, but the model transmissibility.

The algorithm optimizes on the desired number of regions in the simulation model and reduces the complexity of the property matrix size in the AHM process by several orders of magnitude, which is crucial when using CPU time and resource-intensive model-inversion techniques. Fig. 4.14 shows examples of regionalized properties or a reservoir simulation model as a function of iterative progression of the proposed algorithm.

We conclude this section by highlighting E&P applications that use a specific, rapidly emerging field of ML, the so-called deep learning (DL).

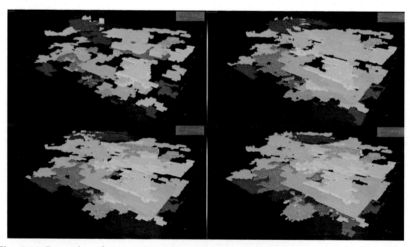

Fig. 4.14 Examples of regionalized properties or reservoir simulation model as a function of expanding random seeds when sampling with the proposed algorithm. *(With permission from Dossary, M., Al-Turki, A., Harbi, B., 2016. Self-Organizing Maps for Regions Exploring and Identification Based on Geological Signatures, Similarities and Anomalies. SPE-182827-MS, https://doi.org/10.2118/182827-MS.)*

The DL approach—also known as the deep structured learning or hierarchical learning—is new and at the forefront of ML research. The main idea is to move ML closer to its AI roots. An example of a DL technique is a deep neural network (DNN), which combines a multilayer network with multiple hidden layers organized into a graph or network (Fig. 4.15), compared with the "classic" NN with a single feed-forward "hidden" layer (see Fig. 4.8A). The main advantage of a DNN over a classic, single-layer NN is the ability to abstract high-level data from extremely complex data sets. For more information on DL and DNN see Goodfellow et al. (2016).

It is encouraging to see that the concepts of DL and DNN have recently started to find their way into E&P research and development, production optimization, and predictive modeling. Crnkovic-Friis and Erlandson (2015) train DNNs to learn the relationships between the geological parameters of nonconventional reservoirs (e.g., thickness, porosity, water saturation, vitrinite reflectance, total organic carbon, brittleness, etc.) and average estimated ultimate recovery (EUR) of an asset.

The DNN model was trained, validated, and tested on a region in the Eagle Ford shale and has included both oil and dry gas wells. The DNN model significantly outperforms both volumetric estimates and type-curve region averages in terms of EUR prediction. However, the most important advantage over traditional decline/type-curve analysis is probably that the DNN model requires geological data only, which means the model can be used in the exploration stage. In contrast, type-curve analysis requires production data to predict EUR, which is only available after a region has been producing for a while.

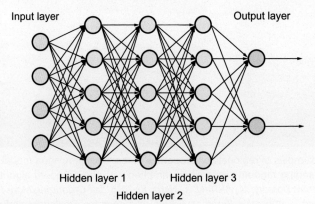

Fig. 4.15 Schematic of a DNN architecture with three hidden layers.

4.3.2 Data Mining, Multivariate, Root-Cause, and Performance Analysis

As a statistical discipline, DM has been used for more than half a century, and has become more applicable and widely used with the emergence of Big Data. Interestingly, one of the fastest tracks in DM development is now seen in the areas of social networking, recommendation systems, and online commerce, which are permanently exposed to enormous volumes of generated data subject to analysis, interpretation, and decision-making (Leskovec et al., 2014; Hallac et al., 2015). In E&P, it seems like more systematic use of DM techniques correlates with the diminishing availability of conventional hydrocarbon resources and the rise of unconventional reservoirs (e.g., shale plays) as the main source of oil and gas.

Numerous publications on the use of DM in the oil and gas industry have emerged in recent years. Moreover, Bravo et al. (2014) have reported that DM ranks among the highest Web-searched term within AIPA technologies. This section briefly summarizes a few recent applications that apply nonlinear multivariate prediction, classification, and root-cause analysis.

Zhong et al. (2015) and Gao and Gao (2013) have cross-evaluated and compared standard univariate linear regression, multivariate adaptive regression splines (MARS) with few more advanced ML techniques, such as SVM, RF, and gradient boosted machine (GBM) to predict the production quality and optimization of almost 500 unconventional wells in both the Permian Basin and Eagle Ford Shale, respectively. Predictor variables include a wide range of categorical and continuous operational and completion well data, such as surface location, architecture (operator, well azimuth, angle, length), stimulation details (fracture fluid, proppant amount, etc.), as well as geological data such as permeability, porosity, viscosity, and other metrics. The production metrics in both studies included a wide range of oil production, accumulated over various periods. To compare the predictive performance of different methods, Zhong et al. (2015) have adopted two objective metrics: the average absolute error (AAE) and mean squared error (MSE). In addition, they have evaluated the tolerances of individual ML methods for missing values, as one of the most common issues in real-world data sets. In terms of overall quality of the predictive fit measured by AAE and MSE, the RF demonstrated the best performance, which is in line with earlier observations explained in Section 4.2.2.

Another frequently used DM technique for multivariate nonlinear predictions is decision trees. For example, Maucec et al. (2015) have deployed classification and regression tree (CART) analysis to investigate whether, from existing well-treatment data, it is possible to find patterns and significant variables that affect the extreme values of pumping job-pause time (JPT) in a particular region, and what is the most critical value causing fracture screen outs. They performed four case studies on a database that included data from 200,000 fracturing and data-acquisition jobs from all over North America, since 2004. The data in the database included: compilation of general well and job information; job-level summary data; pumping schedule stage-level summary data; pumping schedule individual stage data, which included additives, wellbore and completion data, event log data, and equipment data. When mining such complex and extensive databases, the dependencies and correlations among the variables are mostly nonlinear, hidden, and highly nonintuitive. By failing to address such intricate data root-cause relations, frustration can occur when the well operational conditions are thought to be understood but unexpected behavior occurs. This can lead to severe under-performance and economic failure of individual wells, even though the generic data indicates identical formations, similar geologic conditions, and similar completion techniques, as "similar" wells performed significantly better. Similar high dependence on the nonlinear intra-variable effects and potentially negative consequences on the optimization of hydraulic fracturing jobs in shale plays are also reported by Cipolla (2015).

To address the matter Maucec et al. (2015) have built classification trees to predict occurrences of fracture screen outs [as categorical response variables (Fig. 4.16)] and regression trees to predict the JPT as a continuous variable.

They used k-fold cross-validation to assess the misclassification probability for the classification tree and MSE for the regression tree (Fig. 4.17) and performed pruning to optimize the tree depth. In addition, they introduce CART enhancements (Maucec et al., 2012, 2013) by mapping the data points normal (mean $= 0$; variance $= 1$) domain using normal score transform (NST) and kernel k-means clustering to identify the variability of correlated variables and further reduce the sample size. Both enhancements were found to improve the root-cause prediction capability of decision trees by reducing the mean prediction error.

Further examples of predictive modeling with decision trees can be found in Singh (2015) where they are used for root-cause identification and production diagnostics of gas wells with plunger lift and in Schuetter et al. (2015) for production optimization in unconventional reservoirs.

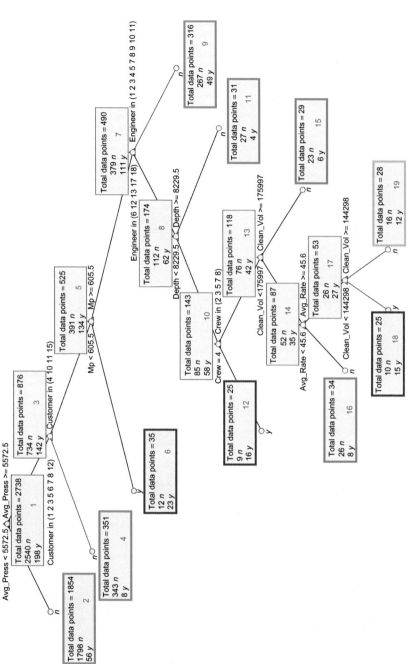

Fig. 4.16 Classification tree for screen out with information for each node given in the text box and numbering given in red. The terminal nodes are colour coded with green (no screen out), yellow (mixed screen out), and red (screen out). Results: 80% = no screen out; 40%–60% = mixed screen out; 40% = screen out. *(With permission from Maucec, M., Singh, A.P., Bhattacharya, S., Yarus, J., Fulton, D., Orth, J., 2015. Multi-variate Analysis and Data Mining of Well Stimulation Data Using Classification and Regression Tree with Enhanced Interpretation and Prediction Capabilities. SPE-166472-PA, https://doi.org/10.2118/166472-PA.)*

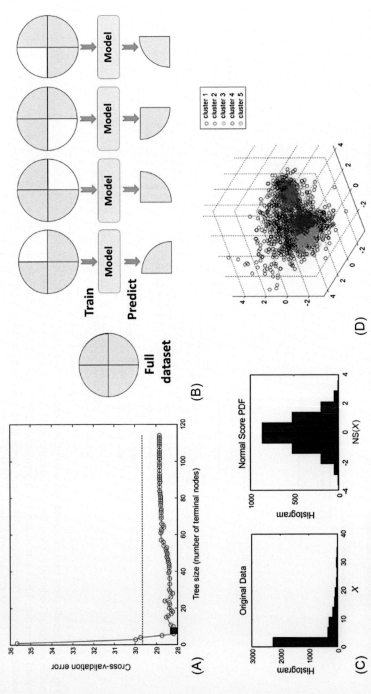

Fig. 4.17 Use of cross-validation in decision tree analysis: (A) Example of cross-validation performed on a regression tree with 113 terminal nodes, with 8 nodes as the optimal point; (B) conceptual representation of *k*-fold cross-validation where *k* = 4; enhancements of CART root-cause prediction capability, (C) NST mapping of JPT variable before building regression tree, and (D) 3D rendering of five clusters after applying kernel *k*-means clustering on JPT data set. *(Parts A and D: with permission from Maucec, M., Singh, A.P., Bhattacharya, S., Yarus, J., Fulton, D., Orth, J., 2015. Multivariate Analysis and Data Mining of Well Stimulation Data Using Classification and Regression Tree with Enhanced Interpretation and Prediction Capabilities. SPE-166472-PA, https://doi.org/10.2118/166472-PA.)*

4.3.3 Event Diagnostics and Failure Analysis

As the applications of Big Data and predictive modeling are rapidly expanding in E&P, they are being used in predictive maintenance, event diagnostics, and failure analysis of a wide range of equipment in oil and gas operations, mainly for drilling and artificial lift (e.g., ESPs) and rotating equipment diagnostics (e.g., compressors and turbines).

The term failure analysis basically stems from the branch of MA, called survival analysis (Tabachnick and Fidell, 2013), which is a set of data-driven statistical techniques (e.g., data mining) for analyzing the length of time until something happens (i.e., an unplanned event) and for determining if that time differs for different groups of samples or for groups subjected to different treatments. For example, in medical settings, survival analysis is used to determine the time course of various medical conditions and whether different modes of treatments produce changes over time. In industries such as oil and gas, such analysis is referred to as failure analysis, and it is used to determine time until failure of a specific equipment part and whether parts manufactured differently have different rates of failure.

Mirani and Samuel (2016) have presented data analytics-based workflows for monitoring and mitigating drill-string failures caused by tool vibrations. The proposed workflow integrates the modified vibration stability plot with the data analytics tool to predict drill-string failures caused by torsional and lateral vibration. In drilling operations, the modified stability plot provides optimum operating parameters—including weight on bit (WOB), revolutions per minute (RPM), and ROP—to minimize vibration. However, the actual real-time generated drilling parameters are not always optimum. Mirani and Samuel (2016) have used the deviation of real-time parameters from optimum values and the tools of unsupervised statistical learning (i.e., data clustering) to calculate deviation vectors, representative of the misfit from the optimum point. The derived stability clusters are then used for quantitative failure mitigation. Moreover, they address the question of, "how long can drilling operations be performed if the measured data remains outside of stability cluster before tool failure occurs." The data-driven calculation of the cumulative vibration risk index provides a more sound technique for risk quantification.

Kale et al. (2015) have proposed another application for optimizing operational performance and failure prevention management of drilling systems using real-time data and predictive analytics. They have proposed the framework and algorithms for constructing data-driven component life models to optimize operational efficiency and extend the life of a drilling system. The objective is to minimize the overall life cycle cost of tools, which includes the cost of maintenance and cost of failure. The optimization variables are maintenance intervals and operational parameters such as RPM, WOB, and ROP. Kale et al. (2015) have integrated qualification test data, operational data, drilling dynamics, and historical FRACAS (Failure reporting analysis and corrective action system) information with mathematical and statistical models—such as a proportional hazard model, cumulative damage model, characteristic life function and maximum likelihood estimation, and outlier detection—to predict the time to failure of critical components. They validated the proposed methods to optimize maintenance intervals of a rotary steerable system with and without a motor.

Popa et al. (2008) have introduced a case-based reasoning (CBR) (Montani and Jain, 2010) approach for well failure diagnostics and planning. The CBR is basically a problem-solving expert system that derives knowledge and expertise from a library of historic cases, rather than from classical encoded rules. The data used in CBR systems usually represents the knowledge, experience, and thought process that the user would exercise in, for example, a well intervention event. Fig. 4.18 shows a diagram of a generic CBR process, adapted from Aamodt and Plaza (1994). Popa et al. (2008) have applied the CBR process to improve sanded/seized well intervention planning. They have demonstrated the significance and unique advantage of CBR tools over other ML methods such as NNs, through efficient

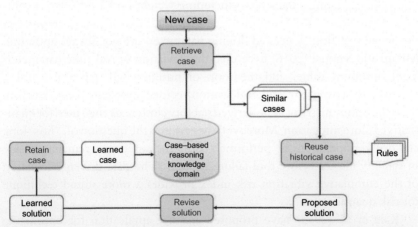

Fig. 4.18 CBR process. Modules highlighted in yellow represent actions in the workflow. *(Modified from Popa, A., Popa, C., Malamma, M., Hicks, J., 2008. Case-Based Reasoning Approach for Well Failure Diagnostics and Planning. SPE 114229, https://doi.org/10.2118/114229-MS.)*

integration of reactive and proactive dynamic data from nearly 5000 well intervention events in a 3-year period, focusing on sanded/seized failure events.

Last but not least, we touch on the application of predictive modeling to fault/failure analytics of ESPs, which are currently the fastest growing artificial-lift pumping technology, deployed across about 15%–20% of oil fields worldwide. However, ESP performance is often observed to decline gradually and reach a point of service interruption because of factors such as high gas volumes, high temperature, and corrosion. Numerous workflows have been implemented to monitor ESP performance and suggest action in case of a failure. However, most such workflows are reactive in nature, where action is taken after the failure event. Recently, the E&P industry has seen an emerging trend of deploying down-hole sensors for real-time surveillance of parameters impacting ESP performance, with an opportunity to predict and prevent ESP failures using data analytics. Such data-driven models would, for example, use the following ESP performance data: historical data with time series values for critical parameters, maintenance logs and calibration data, and operational specifications of the ESPs. Applying analytics to this type of available data provides the ability to rank the ESPs for priority attention based on fault analysis and then recommend appropriate maintenance (repair, rehab, or replace).

For example, Gupta et al. (2015) have proposed a three-stage workflow based on statistical MVA to detect and diagnose impending problems with ESP operation. In the first stage, the key operational variables (decision variables) affecting ESP performance are identified and evaluated. They developed a hybrid monitoring-intervention model based on a robust PCA, which triggers an alarm if the operational attribute under surveillance exceeds the normal operating range predicted by the model. The second stage involves principles of diagnostic analytics (see Fig. 4.2) aiming at the potential cause that led to the failure. To better understand the root cause and take appropriate action, an importance/sensitivity model (see tornado chart in Fig. 4.11) was built to assess the contribution of the various decision variables toward failures and rank them according to these contributions. The third stage of the proposed workflow involves elements of prescriptive analytics (see Fig. 4.2) and suggesting preventive actions. Such data-driven workflows enable building an ESP health monitoring plot (Fig. 4.19), which visualizes the principal components obtained from the model output, capturing observed variances within specific confidence interval limits. Trends or patterns during normal operation are identified and correlated to either

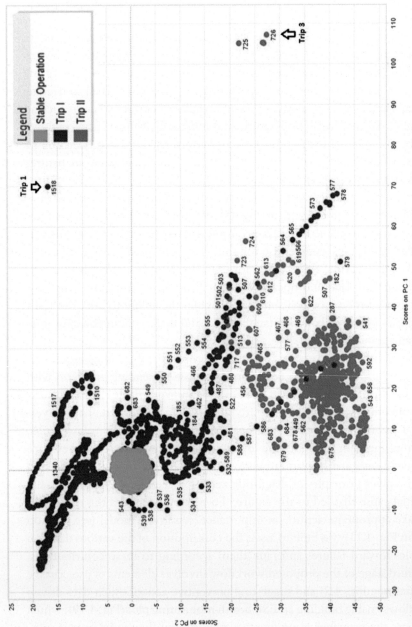

Fig. 4.19 A data analytics-driven workflow can generate an ESP health monitoring plot to predict and identify potential problems before they occur. (With permission from Gupta, S., Nikolaou, M., Saputelli, L., Panjwani, S., 2015 Applying Predictive Analytics to Detect and Diagnose Impending Problems in Electric Submersible Pumps Used for Lifting Oil From Wellbores. Paper 419054, Presented at AIChE Annual Meeting, Salt Lake City, UT, USA, Nov. 8–13, 2015, https://aiche.confex.com/aiche/2015/webprogram/ataglance.html.)

satisfactory operation or observed malfunction. This correlation and knowledge enables a shift toward proactive ESP monitoring to predict and identify potential problems long before they occur thereby reducing intervention costs and optimizing production.

4.3.4 Real-Time Analytics on Streaming Data

This chapter concludes by describing a few trends and applications in the area of data analytics that have the potential to transform E&P in the era of Big Data and high-performance cloud computing. Traditionally, operators have collected massive amounts of data from equipment, fields and assets, and daily operations. Mostly the data are stored and archived, then processed and analyzed as required. Such traditional workflows result in deterministic analytics and passive surveillance. However, SAS (2015), a leading analytics and software company, believes that three technologies will radically overhaul operational capabilities in the oil and gas sector: the IoT, event streaming processing, and prescriptive analytics. Synergistically, these technologies have the potential to rapidly deliver key operational insights and make analytics tools available and impactful "at the point and at the time of decision" by embedding analytics into decision-making and operational processes. The key to implementing ESP technologies lies in enabling contextual and situational analytics "on the fly"—in ultra-speed, ultra-low-latency environments—which creates a technological inflection point for the next-generation remote-control operations. However, we need to note: at this evolutionary stage of fast data and streaming analytics, E&P operations still do not benefit from "real real-time" analytics (i.e., seconds, milliseconds, microseconds), but rather operate at "near real time," approximately several seconds and more—much more when recovering from infrastructure faults.

The concept of stream processing computation or *stream computing* consists of assimilating data readings from the collections of software or hardware sensors in stream form (i.e., as series of sequences or continuous queries), analyzing the data "on the fly," and producing actionable results, preferably in real- or at the right time.

In current E&P practice, one area that benefits most from event streaming processing and prescriptive analytics is drilling operations. Despite the arguments that drilling analytics are not yet oriented toward automatically maintaining the knowledge base through "near real time" updates, Staveley and Thow (2010) have reported that enabling the results in "near real time" is considered to be highly valuable because it provides the capability to develop, monitor, and optimize drilling KPIs—such as cycle time, NPT, and ROP—to lower overall costs and to identify key contributors to rig performance (rig, personnel/process, wired drill pipe, equipment, vendors, etc.). They further emphasized that the additional high value of real-time data in a drilling knowledge base enable current drilling parameters to be displayed next to offset values, along with any other data in a single collaborative workspace, which is updated automatically and in real time and requires minimal manipulation by drilling engineers.

In another paper, Brulé (2013) has introduced a new paradigm for analyzing massive amounts of data, (semi)structured and unstructured, at ultrahigh speeds and frequencies, for Big Data analytics and continuous model updating in E&P. This new paradigm is based on a real-time adaptive analytics and data-flow architecture, which combines stream computing (Fig. 4.20), Hadoop/NoSQL, and Map/Reduce, and massive parallel processing data warehouses (MPP DW). As indicated in Fig. 4.20, stream computing applications are (or can be) represented as data-flow network graphs (Leskovec et al., 2014) composed of *operators*, interconnected by *streams*. The external data feeds can, for example, represent high-resolution imagery, IoT sensor readings, stream of headline news, or market information, such as securities and commodities. The operators implement algorithms for data analysis, such as parsing, filtering, imputation, feature extraction, and classification.

The E&P version of the IoT represents a complex network of sensors and control and automation systems in oil and gas field operations. For example, drilling surveillance, analysis, and optimization vibrational and acoustic data

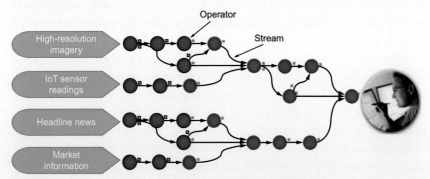

Fig. 4.20 Stream computing applies to high-volume and high-velocity data, whether structured, semi-structured or unstructured. *(Modified from Brulé, M.R., 2013. Big Data in E&P: Real-Time Adaptive Analytics and Data-Flow Architecture. SPE-163721–MS, https://doi.org/10.2118/163721-MS.)*

acquired using emerging sensing technology like distributed acoustic sensors (DAS) or distributed temperature sensors (DTS), which can record events every 1/10 of a second and transmit using down-hole optical couplings. It is worth emphasizing that stream computing does not provide the models needed by E&P. Rather it provides the new concept of computing infrastructure where the data is being generated without concern of scalability, complexity, or bandwidth, and integrated into real-time drilling and production automation and optimization models.

REFERENCES

Aamodt, A., Plaza, E., 1994. Case-based reasoning: Foundational issues, methodological variations and system Approache. Artif. Intell. Commun. 7 (1), 39–52.

Alain Charles Publishing, 2015. Decoding big data, report on big data analytics for oil & gas Conference. Oil Rev. Middle East 18 (7), 160–171. https://issuu.com/alaincharles/docs/orme_7_2015/3.

Begg, S.H., Bratvold, R.B., Welsh, M.B., 2014. Uncertainty vs. Variability: What's the Difference and Why is it Important? SPE-169850-MS. https://doi.org/10.2118/169850-MS.

Braswell, G., 2013. Artificial Intelligence Come of Age in Oil and Gas. SPE-0113-0050-JPT. https://doi.org/10.2118/0113-0050-JPT.

Bravo, C., Saputelli, L., Rivas, F., Pérez, A.G., Nikolaou, M., Zangl, G., et al., 2014. State of the Art of Artificial Intelligence and Predictive Analytics in the E&P Industry: A Technology Survey. SPE 150314-PA. https://doi.org/10.2118/150314-PA.

Brownlee, J., 2014. Jump-Start Machine Learning in R: Apply Machine Learning With R Now. Machine Learning Mastery. http://machinelearningmastery.com/.

Brulé, M., 2009. Using massively parallel processing databases. Digital Energy J. (20), 22–24. http://www.digitalenergyjournal.com/issues/dej20web.pdf.

Brulé, M.R., 2013. Big Data in E&P: Real-Time Adaptive Analytics and Data-Flow Architecture. SPE-163721-MS. https://doi.org/10.2118/163721-MS.

Cipolla, C., 2015. In: How do we optimize hydraulic fracturing in shale resource plays? Learning from the past and predicting the future.Plenary Session at the SPE Hydraulic Fracturing Technology Conference, The Woodlands, TX, USA, 3-5 Feb.

Crnkovic-Friis, L., Erlandson, M., 2015. Geology Driven EUR Prediction Using Deep Learning. SPE-174799-MS. https://doi.org/10.2118/174799-MS.

Davies, B., 2014. Big Data promises big opportunity. Petrol. Rev., 14–16. https://knowledge.energyinst.org/Energy-Matrix/product?product=94368.

Davis, B., 2015. The 7 pillars of Big Data. Petrol. Rev., 34–36. https://knowledge.energyinst.org/Energy-Matrix/product?product=94394.

Dean, J., Ghemawat, S., 2008. MapReduce: simplified data processing on large clusters. Commun. ACM—50th Anniv. Issue: 1958–2008 51 (1), 107–113.

Dossary, M., Al-Turki, A., Harbi, B., 2016. Self-Organizing Maps for Regions Exploring and Identification Based on Geological Signatures, Similarities and Anomalies. SPE-182827-MS. https://doi.org/10.2118/182827-MS.

Fernandez-Delgado, M., Cernadas, E., Barro, S., Amorim, D., 2014. Do we need hundreds of classifiers to solve real world classification problems? J. Mach. Learn. Res. 15, 3133–3181.

Gao, C., Gao, H., 2013. Evaluating Early-time Eagle Ford Well Performance Using Multivariate Adaptive Regression Splines. SPE-166462-MS. https://doi.org/10.2118/166462-MS.

Gelman, A., Hill, J., 2007. Data Analysis Using Regression and Multilevel/Hierarchical Models. Cambridge University Press, New York, NY.

Ghemawat, S., Gobioff, H., Leung, S.-T., 2003. In: The Google file system.Proceedings of SOSP'03 the Nineteenth ACM Symposium on Operating Systems Principles, Bolton Landing, New York, USA, Oct. 19–22, 2003pp. 29–43.

Goodfellow, I., Bengio, Y., Courville, A., 2016. Deep Learning. MIT Press.http://www.deeplearningbook.org.

Gupta, S., Nikolaou, M., Saputelli, L., Panjwani, S., 2015. In: Applying Predictive Analytics to Detect and Diagnose Impending Problems in Electric Submersible Pumps Used for Lifting Oil From Wellbores.Paper 419054, Presented at AIChE Annual Meeting, Salt Lake City, UT, USA, Nov, 8–13. https://aiche.confex.com/aiche/2015/webprogram/ataglance.html.

Hallac, D., Leskovec, J., Boyd, S., 2015. In: Network Lasso: Clustering and Optimization in Large Graphs.Presented at KDD'15, Sydney, NSW, Aug. 10–13, 2015. http://www.kdd.org/kdd2015/.

Handy, A., 2015. Arun Murthy Discusses the Future of Hadoop. SD Times.http://sdtimes.com/arun-murthy-discusses-the-future-of-hadoop/.

Hastie, T., Tibshirani, R., Friedman, J., 2011. The Elements of Statistical Learning: Data Mining, Inference and Prediction. Springer, NY.

Holsman, R., Richards, B., 2016. The 2016 Upstream Oil and Gas Digital Trends urvey, Transcript by Accenture, 16-0274. https://www.accenture.com/us-en/insight-2016-upstream-oil-gas-digital-trends-survey.

Holdaway, K.R., 2014. Harness Oil & Gas Big Data With Analytics: Optimize Exploration and Production with Data-Driven Models. Wiley, Hoboken, NJ. http://www.wiley.com/WileyCDA/WileyTitle/productCd-1118779312.html.

James, G., Witten, D., Hastie, T., Tibshirani, R., 2014. An Introduction to Statistical Learning With Applications in R. Springer, NY.

Jordan, M.I., Mitchell, T.M., 2015. Machine learning: trends, perspectives, and prospects. Science 349 (6245), 255–260.

Kale, D.Z., David, A., Heuermann-Kuehn, L., Fanini, O., 2015. Methodology for Optimizing Operational Performance and Life Management of Drilling Systems Using Real Time-Data and Predictive Analytics. SPE-173419-MS. https://doi.org/10.2118/173419-MS.

Laney, D., 2012. In: Information, Economics, Big Data and the Art of the Possible with Analytics.Presentation by Gartner Inc. https://www-01.ibm.com/events/wwe/grp/grp037.nsf/vLookupPDFs/Gartner_Doug-%20Analytics/$file/Gartner_Doug-%20Analytics.pdf.

Leskovec, J., Rajaraman, A., Ullman, J.D., 2014. Mining of Massive Datasets, second ed. Cambridge University Press, Cambridge, The United Kingdom.

Lochmann, M., Brown, I., 2016. Intelligent Energy: A Strategic Inflection Point. SPE 170630-PA. https://doi.org/10.2118/170630-PA.

Markets and Markets, 2015. Digital Oil Field Market—Global Forecast to 2020. Report Code EP2720.

Maucec, M., Bhattacharya, S., Yarus, J.M., Fulton, D.D., Singh, A.P., 2012. System, Method and Computer Program Product for Multivariate Statistical Validation of Well Treatment and Stimulation Data, PCT Patent Application 2012-IP-061475.

Maucec, M., Singh, A.P., Bhattacharya, S., Yarus, J., Fulton, D., Orth, J., 2013. Multivariate Analysis of Job Pause Time Data Using Classification and Regression Tree and Kernel Clustering. SPE-167399-MS. https://doi.org/10.2118/167399-MS.

Maucec, M., Singh, A.P., Bhattacharya, S., Yarus, J., Fulton, D., Orth, J., 2015. Multivariate Analysis and Data Mining of Well Stimulation Data Using Classification and Regression Tree With Enhanced Interpretation and Prediction Capabilities. SPE-166472-PA. https://doi.org/10.2118/166472-PA.

Mirani, A., Samuel, R., 2016. Mitigating Vibration Induced Drillstring Failures Using Data Analytics: Workflow and case Study. SPE-178849-MS. https://doi.org/10.2118/178849-MS.

Mohaghegh, S.D., 2005. Recent Developments in Application of Artificial Intelligence in Petroleum Engineering. SPE-89033-JPT. https://doi.org/10.2118/89033-JPT.

Mohaghegh, S.D., Abdulla, F., Abdou, M., Gaskari, R., Maysami, M., 2015. Smart Proxy: An Innovative Reservoir Management Tool; Case Study of a Giant Mature Oilfield in the UAE. SPE-177829-MS. https://doi.org/10.2118/177829-MS.

Montani, S., Jain, L.C. (Eds.), 2010. Successful Case-Based Reasoning Applications. Studies in Computational Intelligence 305. Springer-Verlag, Berlin, Heidelberg, Germany, pp. 1–5.

Pokorny, J., 2011. NoSQL Databases: a step to database scalability in web environment. Proceedings of the 13th International Conference on Information Integration and Web-Based Applications and Services, Ho Chi Minhh City, Vietnam, 5-7 Dec. 2011, pp. 278–283.

Popa, A., Popa, C., Malamma, M., Hicks, J., 2008. Case-Based Reasoning Approach for Well Failure Diagnostics and Planning. SPE 114229. https://doi.org/10.2118/114229-MS.

Roth, M., Royer, T., Peebles, R., Roth, M., 2013. In: Using analytics to quantify the value of seismic data for mapping Eagle Ford Sweetspots. URTEC-1619726-MS.Presented at Unconventional Resources Technology Conference, Denver, CO, USA, 12-14 Aug. 2013.

The R Foundation, 2017. Website The R Project for Statistical Computing. https://www.r-project.org/.

Roy, A., Dowdell, B.L., Marfurt, K.J., 2013. Characterizing a Mississippi tripolitic chert reservoir using 3D unsupervised and supervised multiattribute seismic facies analysis: an example from Osage County, Oklahoma. Interpretation 1 (2), 109–124. https://doi.org/10.1190/INT-2013-0023.1.

Saputelli, L., 2015. Transforming E&P Applications Through Big Data Analytics. Webinar, Society of Petroleum Engineers.https://webevents.spe.org/products/transforming-ep-applications-through-big-data-analytics-2#tab-product_tab_overview.

Sarma, P., Leport, F., 2016. Data-Physics: A New Paradigm in Modeling and Optimizations of Oil & Gas Assets. White Paper. http://www.tachyus.com/.

SAS, 2015. How Real-Time Analytics on Streaming Data can Transform the Oil Industry. White Paper. https://www.sas.com/en_us/whitepapers/real-time-analytics-streaming-data-transform-oil-industry-107772.html.

Schuetter, J., Mishra, S., Zhong, M., LaFolette, R.F., 2015. Data Analytics for Production Optimization in Unconventional Reservoirs. URTEC-2167005-MS. https://doi.org/10.15530/URTEC-2015-2167005.

Seltman, H.J., 2015. Experimental Design and Analysis. Chapter, Exploratory Data Analysis. pp. 61–100. http://www.stat.cmu.edu/~hseltman/309/Book/Book.pdf.

Shirangi, M.G., 2012. Applying Machine Learning Algorithms to Oil Reservoir Production Optimization. http://cs229.stanford.edu/proj2012/Shirangi-ApplyingMachineLearning AlgorithmsToOilReservoirProductionOptimization.pdf.

Singh, A., 2015. Root-Cause Identification and Production Diagnostic for Gas Wells With Plunger Lift. SPE-175564-MS. https://doi.org/10.2118/175564-MS.

Staveley, C., Thow, P., 2010. Increasing Drilling Efficiencies Through Improved Collaboration and Analysis of Real-Time and Historical Data. SPE-128722-MS. https://doi.org/10.2118/128722-MS.

Sutton, R., Barto, A., 1998. Reinforcement Learning: An Introduction. MIT Press, Cambridge, MA.

Tabachnick, B.G., Fidell, L.S., 2013. Using Multivariate Statistics, sixth ed. Pearson, London.

Temizel, C., Purwar, S., Abdullayev, A., Urrutia, K., Tiwari, A., Erdogan, S.S., 2015. Efficient Use of Data Analytics in Optimization of Hydraulic Fracturing in Unconventional Reservoirs. SPE-177549-MS. https://doi.org/10.2118/177549-MS.

van Buuren, S., Groothuis-Oudshoorn, K., 2011. Mice: multivariate imputation by chain equations in R. J. Stat. Softw.. 45(3).

Zangl, G., Stundner, M., 2007. Method for History Matching a Simulation Model Using Self Organizing Maps to Generate the Regions in the Simulation Model. US Pat. App. Publication, US 2007/0198234 A1.

Zangl, G., Al-Kinani, A., Stundner, M., 2011. Holistic Workflow for Autonomous History Matching Using Intelligent Agents: A Conceptual Approach. SPE-143842-MS. https://doi.org/10.2118/143842-MS.

Zhong, M., Schuetter, J., Mishra, S., LaFolette, R.F., 2015. Do Data Mining Methods Matter?: A Wolfcamp Shale Case Study. SPE-173334-MS. https://doi.org/10.2118/173334-MS.

Zikopoulos, P., Deroos, D., Parasuraman, K., Deutsch, T., Corrigan, D., Giles, J., 2013. Harness the Power of Big Data: The IBM Big Data Platform. McGraw Hill, NY.

FURTHER READING

Caers, J., 2011. Modeling Uncertainty in the Earth Sciences. John Wiley & Sons Ltd, Chichester, UK.

Doveton, J.H., 1994. Geologic Log Analysis Using Computer Methods. (AAPG Special Volumes).

Magomedov, B., 2006. Self-Organizing Feature Maps (Kohonen Maps). https://www.codeproject.com/Articles/16273/Self-Organizing-Feature-Maps-Kohonen-maps.

CHAPTER FIVE

Workflow Automation and Intelligent Control

Contents

149

5.1 INTRODUCTION TO PROCESS CONTROL

Process control is an engineering mechanism that uses continuous monitoring of an industrial process' operational variables (e.g., temperature, pressure, chemical content) and algorithms and then uses that information to adjust variables to reach product output specifications and objectives. Process control can be a partially or fully automated system capable of maintaining a consistent product output.

Any industrial process loop requires measurement, comparison of data against set points, and continuous adjustments. Systems can be closed feedback or open control. Fig. 5.1 shows a typical feedback system for a fluid tank in the oil industry. Typically, a tank used to store fluids is a closed system where a process variable is measured (process fluids in…), compared to a set point (a maximum allowable level in the tank), and action is taken by a controller to correct deviation from set point. An error signal is generated when the signal value overcomes the set point, and the controller sends a signal to adjust the position of the valve setting (open or close) until the measured fluid has the minimum specification required for this process.

The main components of a control-loop process can be generalized as follows:

- Sensors: electronic or mechanical devices that send signals to transmitters.
- Transmitters: electronic devices that send different types of signals to controllers. A transmitter can send a small current through a set of wires. Signal types are categorized as follows:
  - Analog: continuously varying physical quantity. The most common standard electrical signal is a 4–20 mA current signal.
  - Digital: discrete values that are combined to represent a diagnostic.
  - Pneumatic or differential pressure: using pumped pressure to activate the controller.

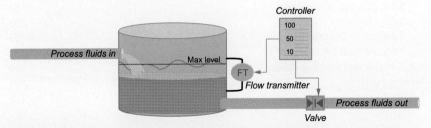

Fig. 5.1 Main components of a control-loop system. In this example, the level of fluid in a petroleum tank is controlled by a valve and level controller.

— Controllers: electronic devices with complex electronic systems and algorithms that operate the controller to act autonomously. Controllers are categorized as

◆ Discrete:

(1) On or off activation, for example, a valve setting that has only two operational positions either open (100%) or closed (0%).

(2) Multiple discrete set points.

— Continuous: automatically compares the value of the input variable to the set point to determine if an error exists. If an error is detected, the controller adjusts its output according to parameters that were set up in the controller.

— Fuzzy controllers: the valve settings follow an adaptive/variable scheme of values. Soft, fuzzy areas of switch control are particularly suitable for systems with high state uncertainty.

To achieve continuous control requires a combination of control modes, which are listed and described below; we believe that continuous control is an essential component of digital oil field (DOF) systems. Table 5.1 compares the advantages and disadvantages of each mode.

Table 5.1 Advantages and Disadvantages of Control Modes

Control Mode	Advantages	Disadvantages
On/off	Inexpensive and simple to install	Operating differential may be outside the process requirement
	Accurate at the point of control	Requires manual intervention
Proportional (P)	Simple and stable	High initial deviation from set point
	Easy to install	Offset occurs near the set point
Integral (I)	No sustained offset	Requires time to overcome instability
	Very precise compared to other control modes	Possible increased overshoot on startup
Derivation (D)	Stable	Requires time to calibrate
	Rapid response to changes; accurate	Some offset
Combination (P + I + D)	Best control	Complex to set up the entire system manually, but electronic controllers allow full automation of a system
	Minimum offset and overshoot	Sometimes expensive

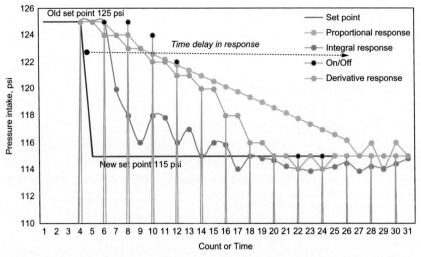

Fig. 5.2 Chart comparing the performance of a valve being operated in four different control modes—proportional, integral, on/off, and derivative—to adjust pressure from 125 to 115 psi, versus response time.

- Proportional (P): the valves and controller are adjusted in different grades based on the change in the measured value from the set point. For example, if a piece of equipment reaches a certain set point, a controller could close a valve by 25%.
- Integral (I): operates at a rate proportional to the magnitude of the input steps.
- Derivative (D): delivers proportional increase or decrease of a variable set as a function of time. Rate action is a function of the change speed.
- Combinations: the best option for oil industry processes and DOF systems. Generally, a combination of P and I is the best option, because integral allows rapid changes on slope and proportional reduces the off-sets. Fig. 5.2 shows a pressure change versus time in response to different control modes.

5.2 PREPARATION OF AUTOMATED WORKFLOWS FOR E&P

Many exploration and production workflows require engineers to coordinate data flows between several different applications. Studies have shown (Al-Jasmi et al., 2013a,b,c) that engineers spend between 50% and 70% of their time gathering, formatting, and translating data to be used

Fig. 5.3 The five main steps to build an automated workflow.

for different engineering applications. Automated workflows not only reduce the valuable time that an engineer must spend doing repetitive, data-preparation tasks, but they also ensure consistency in methods, reduce the probability of input errors, and create a repository of lessons learned and best practices. By definition an automated workflow is a synchronized integration of people, processes, and technology. Fig. 5.3 shows the main steps to build an automated workflow.

5.2.1 Motivation for Automating E&P Workflows

Manual workflows require multiple manual interactions with varied data sources, analytical calculations, and process models. Engineers often work from static electronic forms (e.g., reports, pdf, word processing files and spreadsheets, etc.) that require human data entry and reentry and have multiple related (but offline of the actual data) communications by email, for example, to clarify content or approve next steps. In contrast, an automated workflow integrates client application and/or Web-based dynamic electronic forms, business processes, engineering analytics and modeling, a common data model repository (which can automatically access various data sources), and a self-service workflow application into a comprehensive system that does not require interventions by staff and managers. Fig. 5.4 shows that manual workflows are much less efficient and are prone to errors. Al-Jasmi et al. (2013a,b,c) quantified a comparison between manual processes and automated processes to evaluate and optimize the performance of a well for electric submersible pumps (ESPs) and gas lift (GL), to model the artificial lift and optimize the lift performance. On average, the manual process required 7.2 h per well of an expert's (e.g., senior engineer) time, whereas the automated workflow required 1.6 h per well, by a staff production engineer (PE) (less experienced than an expert) with less risk of data or model errors.

5.2.2 What Kinds of E&P Engineering Processes Should be Automated?

Workflow automation should focus on any tasks that can be done much more efficiently by computers than by people. For example, computing

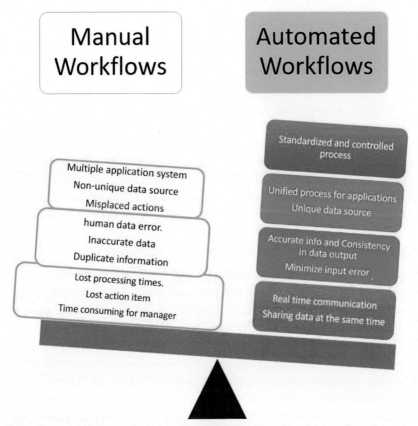

Fig. 5.4 Main challenges of a manual workflow compared to the benefits of an automated workflow.

oil losses and gains for a field with 100 wells or more could require a person to enter values manually into a spreadsheet or database calculator 3 h or more. Alternately, an engineer could use a spreadsheet or other software coupled directly with a historian (which gathers the necessary production data) and uses a batch-allocation process to reduce the total time considerably. However, using cloud services, data analytics, and high-performance computer power, this process can be fully automated and calculated continuously and in real time.

From our collective professional experience, most E&P workflows can be seamlessly automated, particularly those workflows related to production monitoring and surveillance, diagnostic and analysis of production events, production optimization, and forecasting. Automation can be facilitated by a common platform for a data model that consolidates data sources and by a toolkit (commercial or customized technical software) that provides analytic diagnostics and optimization.

Examples of Automated Processes or Candidate Processes for Automation.
Repetitive tasks that require minimum engineering support are as follows:

- *Continuous asset surveillance*:
  - well status (flowing, injection, and shut-in) and production uptime
  - data cleansing and high-level data filtering
  - estimation of production gains and losses against daily and monthly targets
  - production back-allocation
  - real-time KPI production monitoring
  - priority updating of well tests
  - artificial lift status and operational parameters
  - event notification
    - Asset optimization:
      - capacity constraint and operating envelope
      - flow assurance and well integrity

Repetitive tasks that require engineering support are as follows:

- Diagnostics and analysis:
  - filtering and conditioning data using statistical approaches
  - data validation and reconciliation of physical models
  - well rate virtual metering-based model (which includes multiphase flow and well and flow interval allocation)
  - well test validation and well model updates
  - well integrity and failure analysis
  - curve-fitting decline- and type-curve analysis
  - rate and pressure transient analysis
- Optimization:
  - weekly production optimization and short-term forecasts (e.g., 1–7 day ahead) including GL, electro-submersible pump (ESP), and progressive cavity pump (PCP), Rod Pump (RP), and other artificial lift methods
  - monthly production optimization and long-term forecasts (>7 days)
  - production back-allocation and production and injection mass-balance management

Repetitive tasks that require engineering support and include smart analytics are as follows:

- Forecast (prediction) analysis:
    - intelligent alarming by exception-based surveillance
    - event recognition and diagnosis
    - identification and tracking of well production opportunities
    - short-term failure detection of artificial lift (ESP, GL, RP) including early detection of unexpected water production (after water injection)
    - predictive advisory short-term forecasts
    - production optimization and long-term forecasts (>30 days)
    - production and injection management.

5.2.3 Software Components of an E&P Workflow

Automated workflows are a series of processes programmed using a computing language that is capable of executing logical and calculation instructions with minimum human intervention. Saputelli et al. (2013) in a classical paper have showed the best practices in DOF in the last 10 years. They showed a series of technologies which gave origin to the modern integrated analysis. Fig. 5.5 shows the main steps of an automated engineering workflow which include: (1) store information in a database, (2) filter and condition (cleanse) data and extract high-frequency and average data, and (3) send data to the various models and applications in the workflow, which can include both data-driven models and physical models. Fig. 5.5 highlights the importance of these key elements.

Database. Accessibility to data through a database is the starting point of any automated workflow. For production workflows, a database is configured to link data from different data sources (which include data from different time frames and/or frequencies), select and organize data, and send it to technology for filtering and conditioning.

Filtering technology. Mathematical algorithms programmed either in stand-alone or Web-based applications are required to perform data filtering and cleansing (which was discussed in Chapter 3). Many factors, such as signal problems, data transfer, weather/environmental problems, instrument errors, and human interruptions, contribute to errors in the raw data. The filtering process is programmed to remove erroneous data and outliers and provide representative values of the data during the time period of evaluation. When the data is received, cleaned up, and post-processed, the main output is an average value of raw data—that is, production rate (Q), pressure

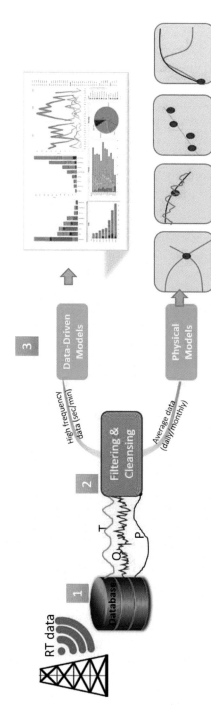

Fig. 5.5 Main components and steps for automated workflow design.

(P), and temperature (T)—at a specified frequency. These values are automatically stored and then sent to various engineering software applications.

Engineer applications. Fig. 5.5 shows the main options used by engineers: data-driven models and physical models. Data-driven models (which were discussed in Chapter 4) are executed using statistics, data mining, data analytics, and application of intelligent components to make decisions, with emphasis on a more qualitative approach. To generate accurately meaningful trends and tendencies, data-driven models need thousands to millions of data points. Physical models, on the other hand, use physical laws and equations that represent production processes and require only a few data points or averages (e.g., daily or monthly data) to generate results.

5.2.4 Modeling the Decision-Making Process

An automated workflow for decision-making should be designed in the same way as any cognitive process; that is, observe, understand (analyze), act, and learn. Bravo et al. (2012) have shown that artificial intelligent component can be the key to enrich the cognitive factor in workflow automation. To assure optimal asset performance, automated workflows include the key phases shown in Fig. 5.6 and described below.

Monitor in real time. Display data time series on a dashboard, with high-definition screens, to show the performance of production over time. Plot and graphical design are set up with a series of rules, with predefined monthly, weekly, or daily targets, and maximum and minimum allowable or target values. Rules are used to estimate absolute differences between actual and target values.

Diagnose and analyze. Once exceeded any maximum or minimum allowable values, the system automatically diagnoses and classifies any events, anomalies, or malfunctions. The workflows can be enriched with fuzzy logic or pattern recognition to allow engineers to be able to differentiate from abnormal situation, equipment failures, or data errors.

Recommend and act. The oil industry has decades of accumulated field experience; thus, engineers know how to act in any specific well issue. Even

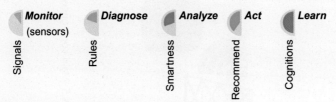

Fig. 5.6 Key phases in a decision-making process.

though each well behaves differently than others, the troubleshooting and remediation process could be unique or slightly different for each situation. Well issues, mechanical malfunctions, electronic equipment failures, or any well anomalies have a recommended action plan. Engineers use supporting tools—such as process models, statistical analysis, data mining technology, expert systems, pattern recognition, neural network tools, physical models, as well as knowledge bases of best practices and lessons learned. The automated workflow should be integrated with these process models to capture knowledge and create rule expert system.

Learn and improve. The final step in an optimal DOF decision-making process is to measure, analyze, and improve the action plan. We believe that all aspects of a decision-making process—monitoring, analysis, actions, and most importantly, the positive or negative results—should be recorded. With current technology, recording the action plan is easy. The challenge is how to process the recorded analysis and results, and then incorporate the learnings to update the automatic decision-making process. Even today no consistent technology exists to capitalize on lessons learned without human intervention.

5.2.5 Automated Workflow Levels of Complexity or Maturity

Brule et al. (2008) describe five levels of workflow automation maturity and where the E&P industry is for each of those levels:

- Level 1: Reporting what has happened: reporting systems, commonplace in E&P.
- Level 2: Analyzing why something happened: ad hoc queries, KPIs, gaining popularity in E&P.
- Level 3: Predicting what will happen or why something might happen: analytical modeling, full-physics models, and integrated asset management (IAM) in E&P (see Chapter 6).
- Level 4: Operationalizing what is happening: continuous update, time-sensitive queries, and in-database analytics on "billions of rows of data—Big Data."
- Level 5: Real-time decision-making to make things happen: actionable data-driven real-time optimization. Early success with event-driven closed-loop IAM in E&P.

Today, many technology companies have reached levels 4 and 5 of maturity; for example, Google, Yahoo, Intel, Facebook, Boeing, Intuit, Amazon, eBay, T-Mobile, ATT, and others. For many large oil and gas companies, levels 4 and 5 are the goals for their DOF automation; however, barriers to achieving these levels include process cost and lack of the right level of expertise in IT to implement the DOF solutions which have hindered digitation in the oil field with only 1% of production data reaching people to make decisions (The Economist, 2017). Unfortunately, today as a whole, the E&P industry is somewhere between levels 1 and 2, with some semipredictive capabilities (level 4) in certain operational areas, such as, gas-lift optimization, ESP, plunger lift, and fracture operations, through the use of full-physics models coupled in a closed-loop IAM optimizer.

On the basis of the degree of smart components (logical solutions without human intervention) and the complexity of the process to integrate and orchestrate data types (categorical, logical, numerical, integer, and string), workflows can be classified as four different types and can be grouped as illustrated in Table 5.2 which shows that for each workflow level, its data frequency, primary tasks, collaboration, and integration with other disciplines, physical model associated with the task, and types of actions:

- manual (100% human intervention)
- semiautomated
- automated:
  - controlled by human operators and
  - autonomous
- smart workflows (up to 20% human intervention; enable decision):
  - self-controlled
  - cognitive and self-trained

Fig. 5.7 depicts a production plot versus time to show the benefit of DOF implementation at different level of complexity of automated workflows. The production profile, depicted with red line represents a manual or semiautomated process (A) without any form of DOF implementation, all the processes are offline, engineers make decisions to change operation settings, the production is affected by excessive downtime and the response time is longer than 24 h. The second system in Fig. 5.7 is the automated but nonoptimized system (B). The process is online and generates rapid diagnostics. The engineers maintain the control, but the behavior of the system is not sustained. The third system is the automated and optimized system (C). The process is fully connected to IOT and coupled to many software applications to allow optimization in real time; however, the operational settings are controlled and supervised by the engineers. The most sophisticated level is the automated, optimized, and self-controlled (autonomous or supervised) system (D), which generates sustainable production gains.

Table 5.2 Level of Complexities of an Automated Workflow

Level Workflow	Data Frequency	Tasks and Activities	Integration	Model	Actions
Manual	Monthly or on demand	Send data through email system or printing report	Production data in silos	Data-info	Reactive with delays. No action, only informing
II Semi-automated (What happened)	Daily to monthly	*Report and describe:* Perform production KPI. Compare actuals to targets Pass data automatically from A to B processes	Production data in silos	Data-info	Reactive with delays. No action, only informing
III.A Automated (controlled; Why did it happen?)	Hour to daily	*Diagnosing and analyzing:* looking for causes why actual production is different from targets. Data is synchronized between A and B	Value chain process, A affects B but B does not affect A	Logical model (if A = B and B > C, then...)	Reactive. Manual actions. Needs approvals
III.B Automated (autonomous, (What will happen?)	Hour to daily	*Predicting:* Predict short term production behavior using statistic on single object data	Production, reservoir and geologist integrated to find causes	Physic model: using app to calculate value	Proactive, using physics to predict 1–7 days ahead
IV.A Smart (self-control)	Real time	*Predicting and Advising:* Predict long term production behavior using multivariable analysis	Losses coupled integration	Physics and data-driven models	Proactive, using physics to predict 1–90 days ahead
IV.B Smart (cognitive)	Real time	*Optimization in* real time	Full integration and closed-loop process (IAM)	Physics and data-driven models	Passive; change control setting before an issue happens

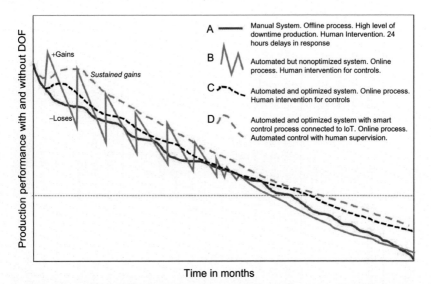

Time in months

Fig. 5.7 Comparison of different levels of automated workflows in DOF. The production benefit of DOF implementation over manual process (A) is observed. There are DOF with automated workflows but nonoptimized (B), automated workflows with optimized system (C) but controlled by engineers and an automated-optimized system and self-controlled by smart operation (D).

5.2.6 The Ten Essential Steps to Build the Back End of an Automated Workflow

In software architecture the back end is defined as a task incorporated as algorithm in a programming script or language which is able to automate the entire workflow process, that is, from A to Z. The back end process also is focused on data transmission, data transformation, data security and accessibility, and backup system. Fig. 5.8 describes the 10 most important steps to design the back end of an automated workflow using a web user interface (UI).

Step 1: Understand the current manual process. Ideally, existing manual processes will be documented. But if not, then it is critical to first document them because it will be very difficult to automate a process until understanding how engineers execute current state process. It is very important to specify the purpose, inputs, outputs/results, dependencies, decisions, and process flow. The most useful way to document a process is a flow chart, which shows objects and actions, and an organigram, which shows

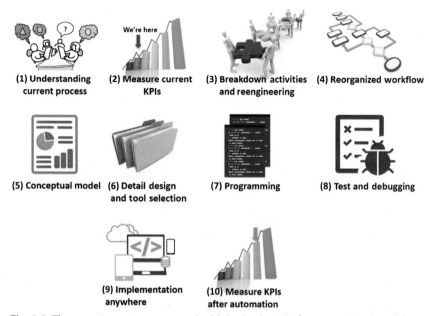

Fig. 5.8 The most important steps to build the back end of an automated workflow.

related organizations that perform the tasks. These graphical tools are not only vital to help identify the critical path, decision points required for the output or the result, but also to help identify problems and bottlenecks, which it will aim to resolve by automation.

Step 2: Quantify and measure KPIs for the manual processes. Measure time, quality, and quantity of the results, accuracy, and repeatability and compare it with KPIs for the desired goals of your organization. (After the process has been automated you might look to improve it by looking at the goals of competitors or world-class benchmark performance.) The KPIs can be visualized using different types of gauges, charts, and bar graphs to show engineers how well the system is doing against defined targets.

Step 3: Break down the activities and prepare to reengineer the whole process. Separate the process into subtasks and set up KPIs for each one. Identify ownership and custodian for each task. Understand the main input in terms of data frequency and logical rules. A swim lane diagram can be useful tool for capturing the data from your analysis. It is important to break down the main activities into subtasks and who (what role or organization) performs each task.

Step 4: Reorganize the workflow. Reorganize the workflow focusing on tasks that deliver business values in terms of engineering time and quality. Avoid redundant activities and subtasks with many approval steps.

Step 5: Build a conceptual model. Prepare a preliminary design in a flow-chart of the workflows, highlighting the main input, engineering process, output, technical applications, and ownership. This is sometimes referred to as functional design.

Step 6: Detail design for the UIs and tool selection. Workflows are made accessible through high-end and visually appealing UIs and user navigation. Web-based workflow is the best technology to generate the visualization for the workflows. To build Web pages, designers can build a series of bluesheets showing the detailed process flow, results, performance metrics, and statistical data enriched with graphical plots, charts, etc.

Step 7: Scripting and programing. There are several technologies available for programming the workflow automatically, sometimes referred to as a "workflow engine." For the authors, the most important criteria to choose the right technology that allow easy, intuitive, and fast interface with the user.

Step 8: Test and debug. This step is one of the most important phases during workflow design and commonly is omitted. In those workflows requesting multiple activities and coupling with several stand-alone applications, comprehensive test protocols followed by debugging will save considerable amount of time and wasted effort ultimately.

Step 9: Implement. The implementation process should assure that the UI can be visualized from multiple devices including PC, mobile phones, laptops, and tablets.

Step 10: Quantify and measure KPIs for the automated processes. Compare the new KPI with step number 2 and show values of improvement.

5.2.7 Foundations of a Smart Workflow

The most effective DOF systems should be designed with advanced automated workflows. Such workflows must be capable of capturing and retaining a company's "knowledge capital" and applying this knowledge to generate faster and more efficient operational solutions. These workflows must also have access to all technical data and technical applications in a unified environment and accelerate repeatable activities based on consistent rules. The difference between an *advanced* and a *smart* workflow is the degree of "cognition" programmed into the workflow and the capacity for reasoning beyond the physical laws that have been programmed into the workflow. Scientists have introduced a series of soft computing techniques (neural networks, fuzzy logic, pattern recognition, etc.) to help workflows improve efficiency and include a level of smartness in the process. However, these techniques require a significant amount of human development and debugging to implement (as discussed in Chapter 4). Soft-computing techniques are difficult to initially implement in automated workflows, but once implemented they enable the workflows to move from merely automated to smart and then truly advanced by doing human functions such as "learn" and acquire "expert" knowledge. Al-Abbasi et al. (2013) and Al-Jasmi et al. (2013a) suggest that the following pillars should be considered during the design of smart workflows:

Knowledge capture. The benefit of knowledge capture is to standardize processes and assure that recurring tasks are done consistently, without fail, and include monthly best practices capable of improving asset performance. Knowledge capture is also important to transfer knowledge from subject matter experts (SMEs) to less experienced engineers. Skin factor increasing with time in oil wells or preliminary values of pump wear factor after 2 years of production are typical examples of knowledge captures. Workflows can be designed to show smart tips that help less experienced personnel understand why data relates to a physical law or that data is out of acceptable range.

Continuous improvement or "Kaizen." A smart workflow will continue to improve and learn if it can recognize patterns in the data and knowledge that it captures. Artificial intelligence systems can recognize patterns in operational variables and predict behaviors, which result in continuous process improvements. Today, technology can generate a self-trained neural network. However, when new events happen, the neural network must be tuned; therefore, there is no complete set of soft computer with a full awareness process.

Multidisciplinary collaboration. Smart workflows should be designed for a high degree of collaboration among disciplines along the E&P value chain, such as subsurface, reservoir, production, operations, drilling, and surface facilities. When working in the system, each discipline should contribute to the main goal and help overcome the complexity of operational problems. Smart workflows use technologies that can support intensive, high-traffic, cross-discipline data flows, information sharing, and knowledge integration.

IT-agnostic solutions. Technology changes so fast that there is often not time to debug an entire workflow. Ideally, smart workflows should be defined independent of the applications or commercial software and systems used to implement them. Al-Abbasi et al. (2013) explains that a well-defined DOF program allows technology to change without significant impact to the structure or performance of the whole solution.

Asset team engagement. As smart workflows are highly automated and can execute quickly, a potential problem can be maintaining people's interest and engagement within the system. Ultimately, the goal of smart workflows is to provide a more effective working environment for asset teams, allowing better communication and analysis to solve operational problems. Al-Abbasi et al. (2013) have described key success factors for keeping people engaged, which includes making sure workflows meet asset team expectations and finding mechanisms to integrate people into daily activities such as daily operational meeting, design of fracture, or drilling the lateral section of a well. New workflows inevitably change the way teams work; all team members must be brought into the program early and should be pretrained to manage the high-stress environment of real-time operations.

Additional details on these topics can be found in Chapter 8.

5.3 VIRTUAL MULTIPHASE FLOW METERING-BASED MODEL

A virtual multiphase flow (virtual flow meter, VFM) metering-based model is an engineering production model that computes multiphase volumes based on pressure and temperature data. Typically, most oil wells (since about the 1990s) are equipped with pressure and temperature gauges to capture frequent readings (e.g., every minute), but production rates are only measured monthly or less frequently with separator tests. Physical multiphase flow meters are very useful and provide significant value to monitor multiphase production in real time, particularly in wells with high production rates where immediately identifying important changes in pressure is crucial. However, these meters are very expensive, and require extensive pressure-volume-temperature (PVT) data and frequent calibration.

Virtual metering is a mathematical model that continuously computes the three-phase flow (water, gas, and oil) rates based on primary real-time data, such as surface and down-hole pressure and temperature, and chokes valve sizes. VFM models are used and reconciled in three zones of operation (beginning at the surface): the wellhead and choke, the well trajectory and wellbore, and the near wellbore-reservoir area.

5.3.1 VFM Physical Models

Choke models. This is the first level of calculation for a VFM model. The choke model is used to determine the flows through a choke or orifice under both critical and subcritical flows. The model uses flow–dynamic equations such as Gilbert (1954), Ros (1960), and Perkins (1993) models to predict and back allocate the total liquid or gas. Eq. (5.1) is a generalized expression of Gilbert correlation as follows:

$$\frac{1}{Q_{L-Surf}} = \frac{a_1 \times THP \times GOR^{a_2}}{\left(\dfrac{d}{64}\right)^{a_3}} \tag{5.1}$$

where Q_{L-Surf} is the total liquid rate at surface condition in STB/d; oil rate can be calculated by multiplying the water cut (WC) ratio with total liquid; *THP* the tubing head pressure in psi; *GOR* the gas–oil ratio in SCF/STB, *d* the current choke size over 64 in.; a_1, a_2, and a_3 are the multiphase flow coefficients taken from different correlations shown in Table 5.3.

Rastoin et al. (1997) have simulated these equations to match rate and observed <13% average error and 17% in standard deviation, when using these expressions under subcritical conditions. They have reported that Perking has the best performance compared to Gilbert (1954) mechanistic models.

Wellbore model. This zone of the system is required to estimate total pressure drop and fluid vertical lift performance (VLP) from the perforation intervals in the wellbore to the wellhead. Starting at the perforation holes and ending at the choke position, the information needed to estimate the VLP are: well trajectories [defined by measured depth (MD) and true vertical depth (TVD)], tubing diameter and roughness, geothermal gradient, GOR, and liquid-gas ration (LGR). Nearly 65% of total reservoir energy losses can occur in the wellbore while lifting the oil to the surface. The best equation to

Table 5.3 Multiphase Flow Coefficients From Various Correlations

Correlation	a_1	a_2	a_3
Gilbert	3.86×10^{-3}	0.546	1.89
Ros	4.26×10^{-3}	0.500	2.00

represent energy loss is by using the relationship between kinetic, gravitational, and viscous/frictional forces as given in Eq. (5.2):

$$\frac{dp}{dL} = \frac{g}{32.2}\rho\sin\theta + \frac{\rho v}{32.2}\frac{dv}{dL} + \frac{f\rho v^2}{2g_c d} \qquad (5.2)$$

where dp/dL is the total pressure losses from a node at a perforation to a node at the surface in psi./ft.; g the gravitational acceleration in ft./s^2; ρ the density of the mixed fluid in lbm/ft^3; θ the wellbore angle; v the fluid velocity across the tubing in ft./s; f the Moody friction factor (dimensionless); d the internal tubing diameter in inches; and dv/dL is the velocity differential per length.

In absence of down-hole pressure gauges, the most important parameter determined by VLP is the flowing bottom-hole pressure (BHP). If we assume a fixed THP, then the BHP is calculated as a function of rate.

Near-wellbore reservoir model. The most common method for modeling this zone is use of the IPR, which calculates the flow rates at bottom-hole, assuming a constant reservoir-pressure-boundary condition. The expression estimates the ability of the reservoir to deliver fluids to the wellbore using Eq. (5.3):

$$PI = \frac{Q_{\text{tot}}}{Pe - f_{\text{BHP}}} \qquad (5.3)$$

where *PI* is productivity index in STB/psi; Q_{tot} the total fluid rates at surface conditions in STB/d; *Pe* the reservoir pressure in psi at the boundary of the reservoir; and f_{BHP} is the flowing BHP in psi at the middle of the perforations.

Numerical and analytical models can predict with high accuracy the well inflow (PI). However, it is time consuming and normally petroleum engineers use well performance software to estimate the PI.

5.3.2 Building Blocks

The IPR and VLP curves can be integrated into a single plot showing f_{BHP} as a function of flow rates (Fig. 5.9). The intersection of the curves indicates the operating point of the current well production in terms of total liquid rate, gas rate, and THP. The IPR/VLP models are calibrated using flow tests at separator conditions approximately every month or quarter; the flow test measures oil and water at a tank and gas using an orifice. With this known production value set point and THP, the f_{BHP} is estimated using the VLP model and extrapolating the down-hole pressure at the perforations. To meet

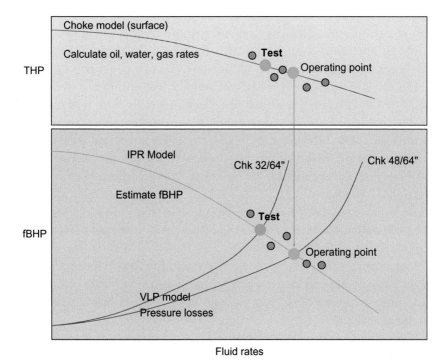

THP

Choke model (surface)

Calculate oil, water, gas rates   ⊙ **Test**

Operating point

IPR Model

Chk 32/64"   Chk 48/64"

Estimate fBHP

fBHP

⊙ **Test**

Operating point

VLP model
Pressure losses

Fluid rates

Fig. 5.9 Construction of a VFM model based on choke, IPR, and VLP models. The plot shows in the top the THP as a function of fluid rate showing both the current operating point of the wells and the value of the test. In the bottom, it shows the extrapolated flowing BHP as a function of fluid rates showing VLPs using different choke size.

at an operating point, the VLP model must match the IPR model. For daily operations and in the absence of well tests, the choke model is used to estimate the flow rate at a given GOR, WC, and THP. The IPR model computes the f_{BHP} based on the rate estimated in the choke model (assuming constant reservoir pressure, WC and GOR from the latest available well test).

5.3.3 Self-Maintaining VFM for a Nonstationary Process

The VFM physical models are based on steady-state condition; it assumes that GOR, reservoir pressure, and WC do not change during the time of evaluation. This calculation is a snapshot of the current behavior or situation. The DOF systems use computational software within a workflow to automatically repeat this task every day, hourly, and even each minute. If the THP changes with respect to the latest THP, the automated workflow can be activated to compute the current production performance. If it is a stationary process, the workflow can run without errors. If reservoir pressure depletes over time and reaches the saturation (fluid bubble point or dew point) pressure, then the situation is more complex. The workflow should be intelligent enough to predict the dew or bubble point pressure and detect significant changes in GOR and WCs that could affect f_{BHP}.

Industry's best practice is to perform periodic well testing to calibrate the values of VFM against the measured collection system. The well tests are usually conducted from skid-mounted separator units, commonly scheduled once every 1–3 months per well. However, in DOF systems, well testing should be scheduled based on an automated well test priority ranking using real-time data and well events to determine if a well should be tested by exception.

An automated workflow is needed to auto-calibrate the VFM, so that the VFM can output production data in real time even if the well is not tested. This workflow is designed to

— Detect automatically the well testing event (using GPS in the mobile tester truck or pressure detection in separator test).
— Capture real-time (24 h) gas, oil, and water volumes, and temperature and pressure data.
— Clean and filter the data from frozen, out-of-range, flagged, and other signal abnormalities.
— Estimate average values during 24 h.
— Feed data into the well performance software.
— Perform model calculations and estimate rates and calculate errors between measured and calculated rate and f_{BHP}.
— Provide historical comparisons with previous well tests and VFM models.
— Provide guidance on whether test data is valid based on model adjustment and errors.
— Adjust models to reduce error, if any, by changing coefficient and time-dependent factors such as PI, skin, and other coefficient correlations, for example, a1, a2, a3, P1, P2, etc., and change stationary data.
— Estimate rate and f_{BHP} using calibrated data up to the next well test.

For example, in an unconventional gas well, the gas flow meter data and gas flow from the VFM were compared to measure the relative error with changes in choke size. Fig. 5.10 shows the daily gas rate reading for a multiphase flow meter, nonstationary VFM, stationary VFM, and well test. The gas flow meter was also compared with well test data, with three changes in choke size, from 24/64 in. to 32/64 in. to 48/64 in. The nonstationary VFM responds quickly to the pressure changes; the regular

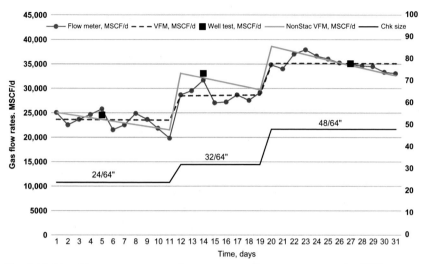

Fig. 5.10 Monthly gas rate comparison among a flow meter, a regular VFM, a non-stationary VFM, and a well test.

VFM (based on constant BHP pressure) just predicts new rates and stays flat until new changes in pressure. The error between the flow meter and the well test is 2%, whereas the error between the regular and nonstationary VFMs is 8% and 6%, respectively.

The relative error of the VFM is calculated in the histogram shown in Fig. 5.11. The statistical distribution shows that 81.5% of total gas readings were in the range of ±8%, which is an acceptable value. The table in Fig. 5.11 shows that the average error could be about ±5; the main reason for mismatch could be attributed to one of the following: (1) reading the wrong THP data (+0.55), (2) adjustment of the multiphase flow correlation coefficient (+0.35), or (3) flowing under loading effect (−0.25).

Using public data we have classified this information for different kinds of hydrocarbon fluids and found that the most complex fluid to be measured with a multiphase flow meter and VFM is gas condensate (error ± 16%), particularly for those wells with f_{BHP} below dew point pressure. One of the possible factors that affect the misreading in gas condensate is the lack of PVT or equation-of-state (EoS) calibrations. In an oil system (±8%), the common error factors are data problems and flowing the well under critical condition. For a heavy oil system, reading gas rate is a common problem (error ± 13%); heavy oil wells produce with slugging flow regimes (difficult to lift oil to the surface) make it difficult for sensors and VFM readings to correctly calculate the values of gas volume.

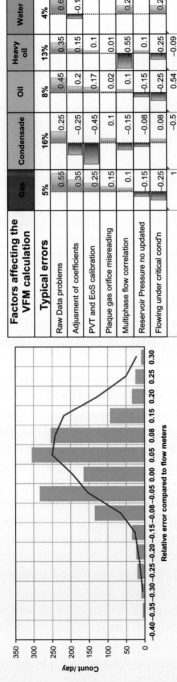

Factors affecting the VFM calculation	Gas	Condensade	Oil	Heavy oil	Water
Typical errors	**5%**	**16%**	**8%**	**13%**	**4%**
Raw Data problems	0.55	0.25	0.45	0.35	0.65
Adjusment of coefficients	0.35	−0.25	0.2	0.15	−0.15
PVT and EoS calibration	0.25	−0.45	0.17	0.1	0
Plaque gas orifice misreading	0.15	0.1	0.02	0.01	0
Multiphase flow correlation	0.1	−0.15	0.1	0.55	0.25
Reservoir Pressure no updated	−0.15	−0.08	−0.15	0.1	0
Flowing under critical cond'n	−0.25	0.08	−0.25	−0.25	0.25
	1	−0.5	0.54	−0.09	1

Fig. 5.11 Relative error distribution in VFM for gas reading compared with gas flow meter. Table on the right shows the common factors affecting the VFM readings per different fluid type.

Fig. 5.12 plots relative errors between a flow meter with a well test (WT) and a VFM with a WT. Fig. 5.12A shows the results of a VFM and a flowmeter, which has a correlation factor of 0.80. These results are because of the highest points observed in the blue square. Those points are attributed to the fact that VFM responds quickly to changes in pressure, whereas, the flow meter measures the flow instabilities during choke changes. Fig. 5.12B shows the VFM relative error to the WT versus the FM relative error to the WT. From this plot, we can make the following conclusions:

- VFM and FM relative errors compared to WTs are positive (quadrant I, Fig. 5.12B). We can infer that the WT is wrong, or both the FM and VFM need calibration. There are points beyond the 15% error radius; both the VFM and FM are reading high rates compared with the WT. These points beyond the error radius are attributed to the rates after changing the choke; however, to avoid ambiguities, the WT must be repeated.
- VFM negative and FM positive (quadrant II). In this quadrant, few points are observed. The VFM needs to calibrate the multiphase flow equation coefficient or PVT data. The plaque orifice needs to be calibrated and the FM needs better tuning.
- VFM and FM are both negative (quadrant III). The WT value is greater than the FM reading and the VFM calculation. The WT must be repeated.
- VFM is positive and FM negative compared to the WT (quadrant IV). Repeat the WT, or calibrate both the FM and VFM. These values are very common when the gas flow is under critical conditions.

5.3.4 Benefits and Disadvantages of Using VFM

If economics allow, it is better to have flow meters on each well. However, VFM offers several benefits such as

- Monitors in real-time multiphase flow gas, oil, and water for all wells in a field.
- Allows operators and engineers to be prepared and react quickly to events, to reduce production downtime.
- Reduces the cost of installing and maintaining multiphase flow meters. Well-tuned VFM software with sufficient data can achieve similar or better results than physical equipment which may drift from calibration.
- Achieves faster and direct production-allocation process compared with typical back-allocation systems which have larger errors.

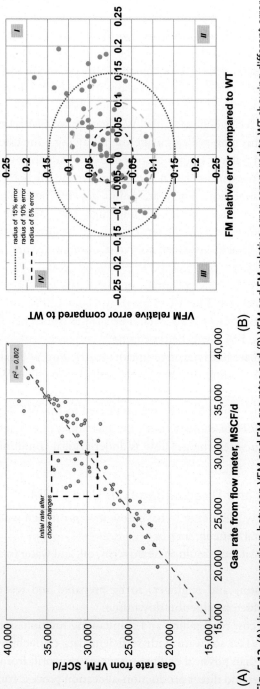

Fig. 5.12 (A) Linear comparison between VFM and FM gas rates and (B) VFM and FM relative error compared to WT showing different error radius of 5%, 10%, and 15%. Gas rate from VFM, WT, and FM is read in MSCF/d.

The disadvantages of using a VFM based on a physical model are that it cannot predict unexpected events, memorize previous events, or learn from previous experience. If changes in pressure, GOR, and water cut are significant, we suggest using an approach that integrates physical and artificial neural network (ANN) models. Use the ANN as a VFM in real time, and use the physical model (updated monthly with test data) to calibrate the ANN.

5.3.5 VFM Based on Artificial Intelligence Models

In KwIDF program (Al-Abbasi et al., 2013), we led a team that designed and applied a series of automated workflows using AI to provide a VFM model, with the ultimate goals of filling in missing data, estimating instantaneous flow rates, and predicting flow rate, WC, and GOR for 7, 15, and 30 days ahead of current production. The work was published in Al-Jasmi et al. (2013b) and Rebeschini et al. (2013).

These automated workflows are built to predict 30 days of production for gas and ESP artificial lift systems. The workflows use a script to collect the previous 90 days of real-time data (7.7E + 6 data points per property) for these properties: THP; ESP pump frequency; casing head pressure (CHP); pump discharge pressure (PDP); pump intake pressure (PIP), motor temperature (MT), and amperage for ESP wells; and CHP, THP, and GL injection volumes for GL wells. The data are stored in a central data base and filtered to prepare average daily calculations. The average values are updated into a well-performance model to train the ANN application subroutine. The ANN uses a radial basis function (RBF) algorithm as an activation function, and the data input in the ANN are saved as text files. The ANN subroutine checks any changes to detect possible well events, such as changes in pump frequency or gas injection volume. The output is used to predict the next day's liquid rate and WC. The challenge was presented when the author tried to predict for 7 + days ahead. To boost the calculation, we introduced Taylor's theorem, for time series regression analysis, to trend backward and forward sampling time; in this particular analysis, we selected −7 and −14 days backward and +7, +14, +21, and +30 days ahead of production data. The Taylor's theorem is given with Eq. (5.4):

$$
\begin{aligned}
f(x) = f(a) + f'(a)(x-a) + \frac{f''(a)}{2!}(x-a)^2 \\
+ \frac{f'''(a)}{3!}(x-a)^3 + \cdots + \frac{f^n(a)}{n!}(x-a)^n + \cdots
\end{aligned}
\tag{5.4}
$$

where a is time (*t*) in days and *x* is *t*–1.

To validate the output of the ANN, the workflow is preset with an internal correlation coefficient (R^2) that rejects and accepts the computed liquid rate or water cut. The automated workflow was proved in the following events or production scenarios:

- Replicate and populate daily production data when pressure, temperature, and production signals are missing.
- Provide ESP production data when data transmission is frozen or electrical power is shut down.
- Generated on–demand sensitivity analyses by changing pump frequency and GL volume.

The prediction results were acceptable out for 20 days (Fig. 5.13). The ANN responds well within acceptable accuracy to changes in THP with pump frequency and THP with gas-lift injection. The ANN was found to be an excellent tool to populate missing data from production history and a useful tool to provide on–demand sensitivities for changes in THP. The ANN proactively predicts the liquid rate and water cut for the next 20 days. However, the ANN cannot predict water cut with acceptable accuracy; the water cut results sometimes appear illogical or do not follow the water-cut history trend. The main reason for this failure is that the incremental water cut is not due to changes in frequency or gas volume but is more related to water injection. We conclude that the ANN is powerful tool to be used as a VFM to measure oil and gas, but should not be relied on to predict oil, gas, and water beyond 20 days. The ANN could predict short-term production (<30 days) with 90% confidence.

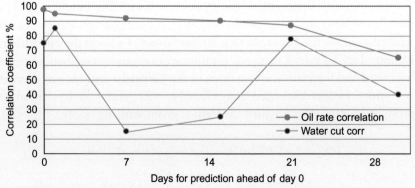

Fig. 5.13 Correlation coefficient for VFM based on an ANN. Oil rate can be predicted with 90% confidence up to 20 days. WC is unpredictable using an ANN.

5.4 SMART PRODUCTION SURVEILLANCE FOR DAILY OPERATIONS

During the last decade, traditional production monitoring has been done using stand-alone applications that require extensive training and a step-by-step processes, which can be tedious and time consuming. These commercial applications display a series of UIs showing Cartesian plots, time series plots, pie/bar graphics, and geographical maps and tables to organize production data. The applications provide excellent solutions for monthly decisions, but in today's DOF we use real-time data. The benefit of using real-time data is to reduce production downtime as much as possible. Two or three days of production losses can mean hundreds of barrels of oil; reducing or preventing production downtime can affect 1%–2% of the total financial impact of a company. Schotanus et al. (2013) have generated a production deferments reports driven by exception-based surveillance process by recommending a series of associated well remedial actions which have resulted in an 8% production gain.

Smart production surveillance is a continuous real-time operation that monitors well surface and down-hole data, helped by predictive tools to foresee upcoming events or unexpected production performance issues, such as early water or gas breakthrough. Smart production surveillance uses a series of UIs enriched with iterative plots, infographic data, maps, and custom layouts that generate actions and recommendations, and pulls up the data required for further analysis. Al-Abbasi et al. (2013) has defined smart production surveillance as an advanced workflow that helps control production and provides surveillance in real time, at various monitoring levels and Al-Jasmi et al. (2013c) developed a series of UIs that allow monitoring, generate alarms, provide diagnostic, and production prediction generation all-in-one.

The main functions of a smart production workflow summarized in Fig. 5.14 are as follows: (1) monitor production data, (2) use filtering and conditioning to calculate average and representative values, (3) calculate the production KPIs compared with targets and goals, (4) generate quick diagnostics based on analytical/numerical models, (5) generate short-term

Fig. 5.14 Main steps of a smart production surveillance workflow.

prediction based on analytical models, and finally (6) generate actions, lessons learned, and recommendations to improve field operation.

5.4.1 Business Model

Smart production surveillance workflows should be built to focus on: (1) controlling, mitigating, and reducing those factors that influence production downtime and total production losses and (2) improving team productivity and process efficiency. The surveillance dashboard enables a management-by-exception approach that significantly improves performance. Main factors that can impact net present value (NPV) and internal rate of return (IRR) include underperforming wells, ESP, GL, PCP, compression, and other facilities that are down, poor data quality, and nonproductive time (NPT) team member. Figs. 5.15 and 5.16 show field-level KPIs and well-level KPIs that can be monitored and improved by smart surveillance to improve business model performance.

5.4.2 Main Components of Smart Production Surveillance

The main components of smart production surveillance include the following:

- *Sensors*: Surface sensors measuring pressure and temperature are required. Production rate or multiphase flow metering are desired, if economics allow; however, a VFM should be an essential tool for each well. Down-hole equipment (ESP/GL/ICV) could help to record flowing bottom-hole pressure and temperature in real time.

Fig. 5.15 Field-level KPIs that contribute to business model performance.

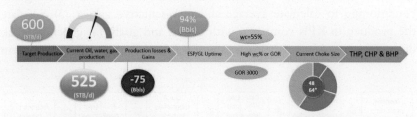

Fig. 5.16 Well-level KPIs used as the main metrics to measure business model performance.

- *Filtering and conditioning algorithms*: It is referred to as doing data preparation using algorithms. Analytical models, such as IPR and Vogel equations, need a single, representative, and average model of the day/week/month or the time period of evaluation. For this, the cleansing algorithms are used to clean data up from data spikes, frozen or missing data, system errors, typos, etc.

- *Well performance model (steady-state condition)*: Examples include the diagnostic model discussed in this section. Any spreadsheet, Web-based, and stand-alone application can be used to compute the flowing condition of a well and estimate the IPR and VLP. This calculation can be executed into the workflow, daily or weekly. The IPR real-time calculation does not have a logical basis because fluctuations in both pressure and flow rate (due to changes in flow regimes, liquid loading, and solid obstructions) can generate misleading results in the IPR calculation.

- *Model-based analytics (data driven)*: The prediction section can be built using an ANN and fuzzy logic techniques to predict 1 to about 30 days of production performance. Alternatively, type curve, DCA, and other curve-fitting methods can help predict production performance with acceptable error and, more importantly, generate immediate actions to prevent high water cut and diagnose underperforming wells.

- Tracking actions and well events using expert systems, pattern recognition, and predictive advisory tools.

5.4.3 UI Dashboard and Layout

The dashboard and layout should include important information affecting the current operations, and it should be concise and well organized so that managers, supervisor, engineers, and technician can easily read and decipher the displayed information. First, display a geographical information system (GIS) with latitude and longitude of well surface locations. This data can be combined with daily production of both water and oil, and displayed as a pie chart. In the background, display the total cumulative oil or gas to mimic the drainage area extrapolated at reservoir conditions with active links to the well and field data, along with metrics on well events including downtime. The dashboard also has queries to alarm and alert management by exception. Fig. 5.17 shows a typical dashboard showing a GIS map with production data and infographic plots. In the middle of the screen, gas, oil, and water production plots versus time are shown with their respective forecasting.

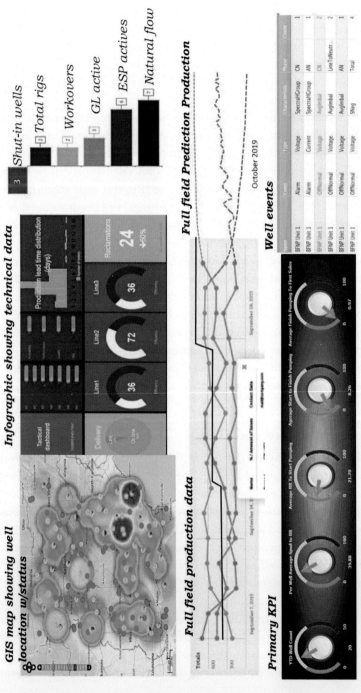

Fig. 5.17 Full field dashboard of a smart workflow showing production prediction and well events.

Well level dashboard. Fig. 5.18 shows a typical dashboard for an individual well presenting a well schematic with real-time data for pressure, temperature and, if available, fluid rate. Connected to this well schematic, the production plot versus the latest 30 days is plotted highlighting the maximum and minimum allowable values (derived from reservoir studies at the maximum bottom-hole drawdown pressure) and the minimum production that satisfies the economic evaluation. KPI indicators are shown at the top right, and the operating point is calculated from the IPR model, with VLP from the VFM. The bottom right displays the forecast derived from the ANN and trained with probabilistic analysis showing three production forecasts (p10, p50, and p90).

5.4.4 What Should Smart Production Surveillance Do?

The following list summarizes the major capabilities and characteristics of a smart production surveillance system:

- Manage a large spectrum of real-time production and pressure data. Approximately manages hundreds of millions of data points in a day (e.g., every minute for 100 wells with 40 tags, each in a field is >150 million points per day).
- Use VFM to compute flow rate for all wells. Delete spikes, frozen, out-of-range data, and populate missing data with physical (logical) information using well performance (nodal analysis) or ANN.
- Manage by exception using smart alarms and alerts generated during the real-time monitoring. Smart alarms filter and rank sensor notifications and well events for redundancy, production impact, and intervention requirements with visual cues so that team members can proactively respond to the most critical production problems in a logical sequence.
- Compute the most important KPIs and other indicators compared to monthly goals and targets. Generate advice based on expert rules.
- Use traffic-light colors as visual cues: red those values in critical condition, yellow values near or over/under the minimum and maximum allowable values, green for those values reaching goals and targets, and blue for those values that are optimized.
- Use up and down arrows to indicate that current production is increasing or decreasing with respect to the last data point.
- The UI should have a highly intuitive design and be interactive enough to allow engineers to introduce their feedback and customize their layout.

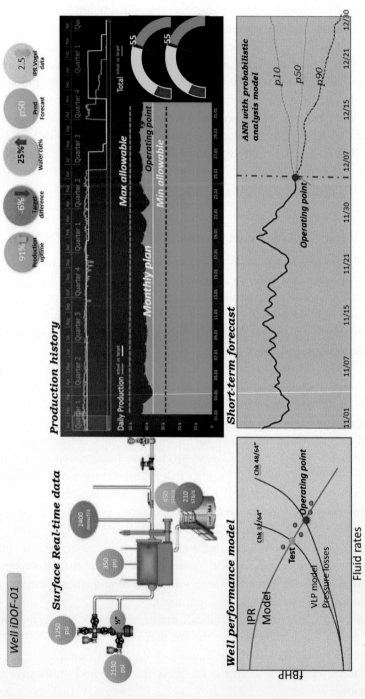

Fig. 5.18 Well dashboard of a smart workflow showing how to integrate real-time KPIs, data analytics, and analytical models in a single layout.

- Use ANNs to predict short-term ($+1$, $+3$, $+5$, $+7$, $+15$, $+30$ days) production forecasts and couple with probabilistic analysis.
- Use fuzzy logic and pattern recognition to predict equipment malfunctions and generate actions to prevent and avoid production downtime for ESP, GL, PCP, RPs, etc.
- Use pattern recognition with expert rules to build a knowledge-capture system and generate solutions for troubleshooting and well remediation.
- Track actions and automatically identify well events, that is, automatically recognize that a well test has started by understanding the pressure buildup without human intervention, or identifying a rig or workover intervention.

5.5 WELL TEST VALIDATION AND PRODUCTION PERFORMANCE IN RIGHT TIME

Well test validation is a crucial component of production performance, particularly for wells without a multiphase flow meter (MPFM) and that depend only on physical models, which must be tuned frequently. If operations and facilities allow it, well tests should be conducted at least once per month. For best results, the test must be performed with three or more choke valve or frequency changes for a period of 24 h or until stability of surface pressure is reached.

The test can be conducted using a portable MPFM (integrated with GPS) or a dedicated flow line test system. The system measures in real-time wellhead pressure, wellhead temperature, fluid rates, choke size, amperage, voltage, frequency, and GL injected volume. With the common RTU system, the signals are sent to a gathering center.

Workflow automation is recommended to collect data, estimate an average value flow rate, update the nodal analysis model, and generate recommendations for the virtual meter. Basically, the workflow should run every day when a flow test is performed. At a minimum, the workflow should perform the following tasks and may also perform others:

- Collect, filter, and clean up the data during the flow periods.
- Automatically identify the flow period by changes in choke settings and measure the response on pressure and rates.
- Automatically identify if the test is running at stable condition ($\delta p/\delta t = 0$).
- Estimate average properties of gas, oil, GOR, water rates, and pressure for each individual flow period with the corresponding 2σ and 3σ standard deviation.

- Transform the wellhead pressure to down–hole pressure using correlated multiphase flow equations (e.g., Duns and Ross).
- Update the well performance analysis (nodal analysis) with average values and build the IPR test. Depending on the model, reservoir pressure, skin, and permeability are known parameters.
- Tune the model-tuning and multiphase flow equations and update.

5.5.1 Key Performance Indicators for Well Tests

Well flowing above critical condition. In gas and gas condensate wells, Turner and Coleman equations are generally used to estimate the critical gas rate (Q_{gc}). The ratio of the current gas rate divided by the critical gas rate (Q_g/Q_{gc}) should be used as a KPI and should be >1.0, which indicates that the well has no issues with liquid loading and production slugging. The test could be performed at a higher rate over the Q_{gc}.

Well flowing at stable condition after each flow. When the pump frequency or choke setting is changed, the changes in pressure over time ($\delta p/\delta t$) should be equal to zero (steady-state flow or a constant (pseudo–steady state). However, when $\delta p/\delta t > 0.0$, the flow is unsteady state and changes to the choke setting should be performed until steady state is reached. Fig. 5.19 shows a prototype of the most important KPIs to monitor a well test in real time.

In 2013, Al-Jasmi et al. (2013d) developed a fully automated workflow that helps engineers analyze well performance with a well test in just 95 min per well, whereas, the typical manual process took up to 5.3 h per well. This change reduced NPT from 5 h/well to just 5 min/well (on average) and increased the time that engineers had to collaborate on more complex issues from a few minutes to 60 min. In this particular multitask workflow,

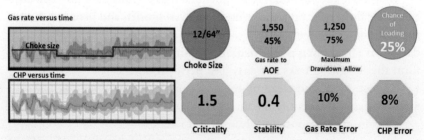

Fig. 5.19 A typical dashboard for a well test validation, which shows gas rate and CHP versus time and choke size. Critical flow passes the check point if the value is >1.0, and stability passes check point if it is >0.0 or is constant with time. Gas rate and CHP errors are computed against the physical models.

the engineers can update the nodal analysis model and compare the result with the previous available test analysis and identify significant changes in IPR or VLP. On the basis of our work on this project, we know that the automated workflow can tune the model parameters until the acceptable error is below 10% between the observed gas rate and f_{BHP}, versus simulated gas rate and f_{BHP}. This workflow displays the multiphase flow equation through time in an iterative GIS map, which shows several wells with different correlations. Ultimately, the physical model is used to validate the well test. If the well test data matches the well model with an error <10%, test is considered validated and accepted; otherwise, the test is rejected and must be repeated.

However, this automated workflow requires the integration of machine learning to memorize previous tuning steps and to be consistent throughout the production history. Normally, the tuning process uses basic equations, such as Vogel, Darcy, or flow parameters such as c and n factors to calibrate the IPR curve. The PEs should have the reservoir pressure, skin, and matrix permeability data; however, if this data is not available, the engineers can change these properties until there is a minimum error, but sometime these changes are meaningless. To avoid meaningless changes, machine learning can be used to memorize the changes in reservoir pressure during the reservoir depletion or provide this value from material balance and numerical models.

The diagnostics provided by the automated workflows need to be reinforced with expert rules analysis, fuzzy logic, and management-by-exception rules. These techniques improve diagnostics of the well troubleshooting and provide accurate recommendations for further action, for example:

- If the stability check KPI is >0.0, the expert rule recommendation could be to reduce the choke size and stop the test.
- If data is frozen, then generate an alarm.
- If WC or GOR increases with a multi-rate test, then suggest the best choke size.
- If gas rate and f_{BHP} do not match, then tune the multiphase flow correlation.

5.6 DIAGNOSTICS AND PROACTIVE WELL OPTIMIZATION WITH A WELL ANALYSIS MODEL

Diagnostics and well optimization are routine activities performed daily by PEs. This section describes typical diagnostics and procedures for optimizing well operations during the construction of an automated production workflow.

Diagnostic actions use the best average surface production and pressure data (from the well test data described in Section 5.4), transform the surface pressure to down-hole data, and then estimate the well productivity index relationship (IPR). The IPR curve is the main tool that a PE uses to evaluate well deliverability. Next, the PE must understand liquid lift performance; the choke setting, GOR, tubing size, depth of end of tubing, and frequency are the preliminary data needed to build the multiphase flow equation and finally the VLP curve. The diagnostic and troubleshooting can be classified depending on if the well is operating under natural flow or the artificial lift method used such as GL, ESP, or PCP.

5.6.1 Natural Flow

Input: The essential data for natural flow diagnostics are THP, choke size, and water, oil, and gas meters. In the absence of an MPFM, VFM validated periodically with well tests are required to evaluate the diagnostic in natural flow wells.

Constraints: Minimum allowable f_{BHP}, pressure drop, maximum velocity to avoid early water breakthrough or sand screen out, maximum rate to avoid conning or cresting, and maximum f_{BHP} to avoid fracture pressure.

KPI: Typical KPIs used for diagnostic are: downtime production, difference with maximum allowable.

Output: Productivity index values (IPR), skin factor, AOF, $k*h$, and estimation of reservoir pressure.

Control: Choke size.

Typical diagnostic:

- Pressure drop increasing due to fast reservoir depletion or additional skin generated by solids production.
- Well flowing under slugging state due to flow regime changing to critical condition.
- Oil rate drops due to early water breakthrough.

5.6.2 ESP and PCP Systems

A modern ESP system is equipped with panel control and a variable speed drive (VSD), which is connected down-hole to the well through electrical cables with sensors capable of reading discharge pressure (PDP), intake pressure (PIP), motor temperature (MT), and many other operating data. These components are shown in Fig. 5.20.

Input: WHP, THP, water, oil, and gas meters. Frequency, amperage, voltage, and choke size are the traditional surface data observed in the control panel. PIP, PDP, and MT are the down-hole data read directly

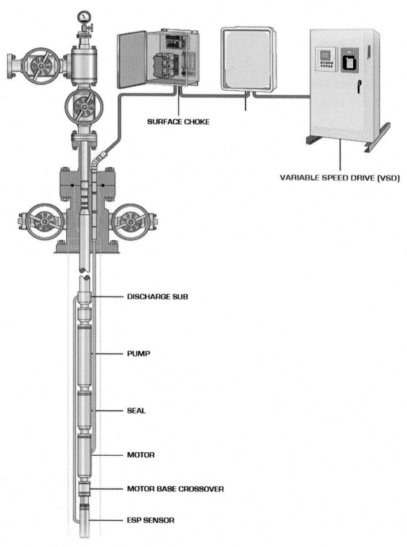

SURFACE CHOKE

VARIABLE SPEED DRIVE (VSD)

DISCHARGE SUB

PUMP

SEAL

MOTOR

MOTOR BASE CROSSOVER

ESP SENSOR

Fig. 5.20 ESP diagram showing the most important components useful for monitoring in real time.

in the control panel. In the absence of an MPFM, VFM validated periodically with a well test are required to evaluate the diagnostic in wells with ESP or PCP systems.

Constraints: Minimum allowable f_{BHP}, pressure drop, maximum velocity to avoid early water breakthrough or sand screen out, maximum rate to avoid coning or cresting, maximum f_{BHP} to avoid fracture pressure, and maximum operating frequency.

KPI: Typical KPIs used for diagnostic are: pump wear factor, gas interference, chance of tubing leak, viscous effect in pump, and solid plugged intake.

Output: Pump head, liquid rate, pressure drop in pump, and pressure drop in reservoir.

Control: Pump frequency and choke size.

5.6.3 Diagnostic Procedure

The value of diagnostics in real time is to prevent additional workovers or well interventions. Real-time diagnostics could be the most effective process to identify pump degradation and impairment in the life of a well. The full diagnostic process should include the following processes:

Alarm system: Generate a primary data set with minimum, maximum, and average values and monitor in real time values that exceed the threshold ranges. Validate if monitoring values are persistent in the next 24 h and apply filtering algorithms to clean false data.

Diagnostics based on a nodal analysis model: The average production and pump data are sent online to a preexisting well model, which is calibrated with the latest well test information, pump design, well trajectory, PVT data, and completion schematic.

Automated model analysis: The model is updated with the latest 24-h (week, month) average of real-time data, such as pressure and temperature data, THP, THT, PIP, PDP, PIT MT, flow data, WC, gas rate, and oil rate. ESP model tuning is performed online daily by comparing the model and sensor data.

Model match: Fig. 5.21 shows a typical ESP pump gradient showing f_{BHP}, PIP, PDP, and THP for a well model and sensors. The model computes different values between the calculated and the modelled data. The tuning process is matched by changing pump wear factor, multiphase correlation friction factors, multiphase correlation gravitational factor, and productivity index. In cases where the system cannot find a solution

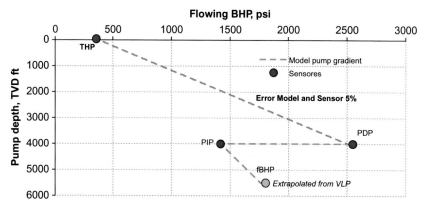

Fig. 5.21 ESP gradient plot showing in real time three sensors (THP, PDP, and PIP) and a model plot (shown in a line) showing the error between the model and sensors. Extrapolated f_{BHP} is also shown.

(unmatched process), a quick diagnostic should be provided, for example:

- Misfit between VFM and test rate is high. Suggestion: check the productivity index or conduct a buildup test to evaluate the current reservoir pressure and skin factor.
- Misfit between PDP-PIP$_{Model}$ and PDP-PIP$_{Sensor}$. Review the wear factor of the pump. Check changes with fluid viscosity and API.
- Misfit between calculated and extrapolated PIP and f_{BHP}. Calibrate the multiphase flow correlation, friction, and gravitational parameters.

5.6.4 Smart Diagnostics

The traditional ESP/PCP diagnostics in real time uses a physical model to evaluate the range of operations that are commonly designed for steady-state condition, where reservoir pressure is not changing. A smart diagnostic not only uses artificial components such as fuzzy logic and neural networks to predict in advance any ESP/PCP troubleshooting or malfunctioning of pumps, but also uses field statistical data and expert rules to generate optimization in real time. Working with Al-Jasmi et al. (2013e), a fuzzy logic algorithm was created to predict pump malfunction for 7, 15, 30, and 90 days ahead of current production. The fuzzy logic was combined with expert rule to diagnose and rank the following pump conditions:

- pump wear factor,
- solid plugged intake,
- gas interference and blocking,

- tubing leak,
- viscous fluids friction,
- electrical issue.

They linked these conditions to the signal of the pump using the matrix illustrated in Table 5.4.

By definition, fuzzy logic techniques assume that all the conditions could occur without exception but in different levels or degree. However, to calibrate the fuzzy logic, ESP experts use weight factors to assign important contributions of developing signs: it is to alert that a pump malfunctioning is in progress. In other words, the ESP experts assign arbitrary but relative weight values per condition. For example:

Signal Condition	Intake Pressure	Current Amperage	Motor Temperature	Pressure Discharge	Liquid Rate	Total
Gas interference	0.2	0.25	0.1	0.25	0.2	1.0

The expert considers that if signals such as intake pressure, current amperage, pressure discharge, and liquid rate decrease and MT increases, it is inferred that "gas interference" condition can occur with a 20% chance affected by intake pressure, 25% chance affected by current amperage, 10% chance by MT, 25% by pressure discharge, and 20% by liquid rate. The results are multiplied by the expected ± signal. If a signal does not behave as expected, the value is rejected. It only sums those signals that follow the pattern. The calculation is repeated for other conditions at different times. The condition that gets the maximum index closest to 1.0 or larger than 0.63 is used as the most likely index, between 0.33 and 0.63 is used as a possible condition, and between 0.0 and 0.33 is assigned as unlikely and discarded from the action plan. Table 5.5 shows an example of a calculation for gas interference.

Table 5.4 Matrix to Link Signal Conditions to Contributing Factors

Signal condition	Intake pressure	Current amperage	Motor temperature	Pressure discharge	Liquid rate	Code
Pump wear	+	-	-	-	-	10000
Plugged Intake	+	-	+	-	-	10100
Gas Interference	-	-	+	-	-	00100
Tubing leak	+	-	Don't include	-	-	1000
Viscous fluid	Don't include	+	+	-	-	1100
Electrical Issue	+	-	+	-	+	10101

Table 5.5 Example of a Calculation for Gas Interference

Signal condition	Intake pressure	Current amperage	Motor temperature	Pressure discharge	Liquid rate	Code
Actual signal	-	-	+	+	-	00110
Preset signal for Gas interference	-	-	+	-	-	00100
Preset Weighting factor for Gas interference	0.2	0.25	0.1	0.25	0.2	1.0
Expected signal	Yes	Yes	Yes	No	Yes	
Calculation=>	0.2	0.25	0.1	x	0.2	0.75

The results are shown in a real-time dashboard in the DOF operation center accompanied by suggested actions. For example, if the most likely event is gas interference, the suggested action is to reduce the choke size to increase the f_{BHP}.

5.6.5 Artificial Lift Optimization

The previous sections have covered the surveillance and diagnostics for VFM. The automated workflow utilizes those to provide advisories on new operating conditions to optimize the operations for optimal production. The well performance models are built in many commercial software applications and the models are calibrated with production tests. The optimization model should first be adjusted for previous cases. Generally, well performance commercial software has interactive functions that allow performing history matches and sensitivity analyses with different operating points for choke size, pump frequency, line pressure, etc. The general steps to generate an optimization to improve the production of ESP as an example are as follows:

- *Model the ESP actual operating point.* We recommend using an average of the last 24 h of production data (pressure, production, and temperature).
- *Verify if the electrical power and electrical current conditions are working properly.* Determine if the pump can handle more energy consumption, if the motor can generate the required power to run the pump, and if the cable is designed to provide electrical current to the motor.
- *Evaluate the actual condition of production and pressure, and plot the operating point of the pump envelop.* Fig. 5.21 shows the ESP envelope for a well with 3–1/2 in. OD tubing size. Check if the operating point is above the minimum operating point or below the maximum operating point.

• *Optimize in real time.* The well performance model is run for many cases, automatically changing both frequency and choke size (if setup with a choke setting). All the cases are run for the available power and electrical current condition (current stable). The most likely points are plotted within the pump envelope (the rest are discarded). The optimum point is highlighted in blue to show the point that is most efficient, shortest, and consumes the least energy.

Fig. 5.22 shows pump head (ft) versus total liquid rates (STB/d) for tubing size of 3–1/2 in. The envelope is limited by the pump maximum and minimum operating lines (in red) and an efficiency line (green). Frequency lines are observed from 40 to 60 Hz. The chart shows 10 points of possible scenarios where production can increase. However, the algorithm is designed to find the shortest path, with less power consumption, and production increase without decreasing the f_{BHP}. The path shown in Fig. 5.22 shows two segments: (1) decrease f_{BHP} by reducing the choke size and (2) increase the pump frequency from 42 to 46 Hz to increase production up to 620 STB/d.

Other artificial lift systems, for example, RP, PCP, or GL have analogous optimization procedures supported by DOF systems.

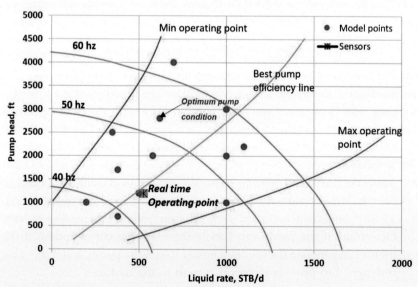

Fig. 5.22 ESP head plot showing the ESP envelope for 40, 50, and 60 Hz of running operation. The real-time operating point is running near the efficiency pump line and computed optimum point (blue). >10 points (red) are highlighted as possible changes to increase production or improve pump efficiency.

5.7 ADVISORY AND TRACKING ACTIONS

Improved accountability is one of the most important components to improve capital efficiency and reduce cycle time. Logging events and recording actions in a database is an essential component to more effective workflows, that is, tracking actions. The insight here is to measure and record input, actions, and outcomes. However, oil companies (like other large complex companies) struggle to manage many data sources, multiple applications, duplication of information, and many layers of communications. To avoid this system complexity, DOF systems are enriched with dashboards to display alarms, alerts, and a list of tickets performed by other engineering workflows. A comprehensive model published in Al-Jasmi et al. (2013f) shows the basis of a fully automated, closed-loop integrated workflow in which, after the proper personnel approvals, the system is able to automatically record the operational changes.

On the basis of our own experience, we believe that the best way for operations personnel to manage actions is by opening and closing tickets in a managed system environment. The ticket can be generated using a well ID and ranked by priority and a ticket's current status. Depending on the level of authority and degree of empowerment, team members can be organized, for example, as monitoring team, engineering team, and operation team. It is critical that all well operational events, changes, and interventions be recorded in a managed data environment.

The main steps of a tracking action workflow can be summarized as follows:
1. Collect and manage alarms and alerts. Discard alarms that do not exceed the thresholds.
2. Classify and group alarms and alerts depending on the volume of production lost compared to the plan, oil or gas deferral, water increase, GOR increase, approaching the critical gas rate, or critical drawdown.
3. Generate ticket and assign team responsibility with follow-up actions. Categorize the tickets by well ID and order of importance for the organization.
4. Submit the ticket to the engineering team to evaluate if production underperforms using physical models. Use the workflow to generate diagnostics and production opportunities.
5. When the engineering team finds an issue or opportunities, the ticket is transmitted to the operation team to execute actions.
6. The monitoring team closes and stores the tickets.

7. Supervisors and managers review the ticket status KPIs on their dashboard.

Ticketing and flow process:

1. The monitoring team ranks the number of issues, alarms, and alerts and decides to generate tickets with comments and follow-up actions.

2. The engineering team receives an email or alert on their mobile app with actions to diagnose and evaluate the current performance of the well. Provide recommendations and assign tasks to the operation team. The monitoring team receives the ticket back and sends it to the operation team.

3. The operation team completes the activity by performing a well operation, such as changing the ESP frequency, choke size, valve for GL injection, and separation system line or RP condition. The monitoring team receives an automated notification that the changes have been executed and the operations team verifies.

4. The monitoring team:
   - Verifies the changes of the physical settings and storage of the events in data system.
   - Verifies that changes in surface pressure, rates, and temperature have been observed (in case of any unresponsive action).
   - Confirms with the operation team that action has been executed.

5. The monitoring team closes the tickets.

REFERENCES

Al-Abbasi, A., Al-Jasmi, A., Goel, H.K., Nasr, H., Cullick, A., Rodriguez, J.A., Carvajal, G.A., et al., 2013. New Generation of Petroleum Workflow Automation: Philosophy and Practice. SPE-163812-MS. https://doi.org/10.2118/163812-MS.

Al-Jasmi, A., Goel, H.K., Rodriguez, J.A., Carvajal, G.A., Velasquez, G., Scott, M., et al., 2013d. Maximizing the Value of Real-Time Operations for Diagnostic and Optimization at the Right Time. SPE-163696-MS. https://doi.org/10.2118/163696-MS.

Al-Jasmi, A., Al-Zaabi, H., Goel, H.K., Lopez, C., Carvajal, G., Vanish, D., Querales, M., Villamizar, M., Yong, Y., Rodriguez, J., Mahajan, A., 2013a. A Real-time Automated "Smart Flow" to Prioritize, Validate and Model Production Well Testing. SPE-163813-MS. https://doi.org/10.2118/163813-MS.

Al-Jasmi, A., Goel, H.K., Nasr, H., Carvajal, G.A., Johnson, W., Cullick, A.S., Rodriguez, J.A., Moricca, G., Velasquez, G., et al., 2013b. A Surveillance "Smart Flow" for Intelligent Digital Production Operations. SPE-163697-MS. https://doi.org/10.2118/163697-MS.

Al-Jasmi, A., Nasr, H., Goel, H.K., Moricca, G., Carvajal, G., Dhar, J., Querales, M., Villamizar, M., Yong, Y., Rodriguez, J.A., Bermudez, F., Kain, J., 2013e. ESP "Smart Flow" Integrates Quality and Control Data for Diagnostic and Optimization in Real Time. SPE-163809-MS. https://doi.org/10.2118/163809-MS.

Al-Jasmi, A., Goel, H.K., Nasr, H., Querales, M., Rebeschini, J., Carvajal, G.A., et al., 2013c. Short-Term Production Prediction in Real Time Using Intelligent Techniques. SPE-164813-MS. https://doi.org/10.2118/164813-MS.

Al-Jasmi, A., Nasr, H., Goes, H.K., Vilamizar, M., Carvajal, G.A., Vellanki, R., Cullick, A.S., Rodriguez, J.A., Raghavendra, N., Dunbar, A., Velasquez, G., 2013f. An Automated "Smart Flow" for Tracking and Assigning Operational Accountabilities in Real Time. SPE-163810-MS. https://doi.org/10.2118/163810-MS.

Bravo, C., Saputelli, L., Rivas, F., Perez, A., Nikolau, M., Zangl, G., et al., 2012. State-of-the-Art Application of Artificial Intelligence and Predictive Analytics in the E&P Industry: A Technology Survey. SPE-150314-MS. https://doi.org/10.2118/150314-MS.

Brule, M., Charalambous, Y., Crawford, M.L., and Crawley, C. 2008. Bridging the Gap between Real-Time Optimization and Information-Based Technologies. SPE-116758-MS. https://doi.org/10.2118/116758-MS.

Schotanus, D., Cramer, R., Babbar, S., Brouwer, R., 2013. Real Time Surveillance and Optimization of a Heavy Oil Field. SPE-165475-MS. https://doi.org/10.2118/165475-MS.

Gilbert, W.E., 1954. Flowing and Gas-Lift Well Performance. Drilling & Production Practice. vol. 13. API, Dallas, TX, pp. 126–157.

Perkins, T.K., 1993. Critical and Subcritical Flow of Multiphase Mixtures Through Chokes. SPE-20633-PA. https://doi.org/10.2118/20633-PA.

Rastoin, S., Schmidt, Z., Doty, D.R., 1997. A Review of Multiphase Flow Through Chokes. https://doi.org/10.1115/1.2794216.

Rebeschini, J., Querales, M., Carvajal, G.A., Villamizar, M., Md Adnan, F., Rodriguez, J., et al., 2013. Building Neural-Network-Based Models Using Nodal and Time-SeriesAnalysis for Short-Term Production Forecasting. SPE-167393-MS. https://doi.org/10.2118/167393-MS.

Ros, N.C., 1960. An analysis of critical simultaneous gas/liquid flow through a restriction and its application to flow metering. Appl. Sci. Res. 9 (Series A), 374.

Saputelli, L.A., Bravo, C., Nikolaou, M., Lopez, C., Cramer, R., Mochizuki, S., et al., 2013. Best Practices and Lessons learned after 10 Years of Digital Oilfield (DOF) Implementations. SPE-167269-MS. https://doi.org/10.2118/167269-MS.

The Economist, 2017. Oil Struggles to Enter the Digital Age. https://www.economist.com/news/business/21720338-talk-digital-oil-rig-may-be-bit-premature-oil-struggles-enter-digital-age.

FURTHER READING

Cramer, R., Jakeman, S., Berendschot, L., 2006. Well Test Optimization and Automation. SPE-99971-MS. https://doi.org/10.2118/99971-MS.

CHAPTER SIX

Integrated Asset Management and Optimization Workflows

Contents

Digital oil field (DOF) systems conventionally have focused on wells, production, and operations. However, DOF is expanding its footprint into field decisions and management. Thus, to optimize production and recovery, production system models are increasingly being integrated with reservoir models. DOF systems now deploy three-dimensional (3D)-coupled subsurface and surface models that, when calibrated (i.e., history matched), provide short- and long-term forecasts of asset production and performance. The main objective of this chapter is to give an overview of the modern integrated asset modeling (IAMod) practices and outline techniques and workflows for optimization and decision-driven forecasts of DOF systems that is integrated asset management (IAM), including model and data uncertainty. In Section 1.5 of Chapter 1, we introduced the concept of optimization process in DOF. The optimization process is related to the area of real-time production optimization for artificial lift (GL, ESP, PCP, etc.) and also applied to maximize the company indicators such as NPV, IRR,

Intelligent Digital Oil and Gas Fields
https://doi.org/10.1016/B978-0-12-804642-5.00006-2

or recovery factor. In this chapter, we discuss the techniques applied for history matching the production data and an overview on coupling subsurface and surface models.

This chapter introduces the engineering principles and technology concepts of IAMod, and optimization and the use of the models for IAM. The IAMod developments were introduced in the E&P industry in the early 2000s (Liao and Stein, 2002), combining subsurface models with surface production facilities and networks. They are now becoming a standard means of modeling entire oil and gas assets with the objective to optimize existing facilities and to plan enhancements to production (wells and facilities).

To date, many operators have developed their own IAMods (Toby, 2014 and references therein); however, the ultimate benefits of such models and particularly their business-added value have not been fully realized. The underlying reasons may stem from the inherent complexity of the state-of-the-art IA models, which makes for challenging model calibration (history matching) that leads to uncertainty in the models and their forecasts (Maucec et al., 2011). Thus, there is an added challenge that is attributed to the fact that the modern IAMods now integrate advanced uncertainty and risk management (URM) principles, which—particularly when deployed in large-scale, full-field studies—require substantial computational resources (Dzuyba et al., 2012; Maucec et al., 2017). While the integration of URM concepts in IAM processes enhances their operational applicability and excellence, it also adds to the challenge of quantitatively evaluating various development and economic scenarios. Moreover, the integration of URM and IAMod ultimately leads to integrated asset management—a decision-driven reservoir optimization, which in open technical literature is interchangeably (and sometimes confusingly) associated with the same acronym IAM.

6.1 INTRODUCTION TO IAM AND OPTIMIZATION

Before introducing the principles of IAMod and optimization, let us look briefly at the notation and abbreviation patterns that one will encounter in the open technical literature, for example, in the publications of the Society of Petroleum Engineers (SPE).

In this work, the integrated asset model or modeling is "IAMod," while integrated asset management is "IAM". This notation is incorporated in Fig. 6.1, which outlines the framework for a holistic IAM. The reservoir

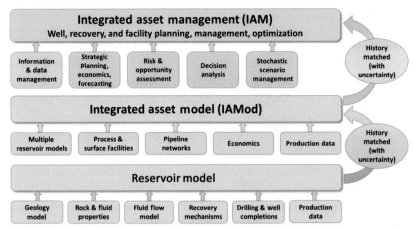

Fig. 6.1 Holistic concept of integrated asset management (IAM).

simulation model represents a starting point and one of the input modules for an IAMod. A reservoir model integrates the reservoir geological model, rock and fluid properties, fluid flow model with recovery mechanisms, drilling and well completions, and the production history. The concept of IAMod introduces multiple model realizations as a result of uncertainty quantification and modeling, integration with process and surface facilities, and pipeline networks and economic exploitation models.

With integration of information and data management, strategic planning with risk and opportunity assessment and decision analysis—all in the frameworks of stochastic scenario management and optimization—IAM concepts are being used more often. They are now becoming a standard way of modeling entire oil and gas facilities with a view to optimizing existing facilities or developing new ones, as well as applying them to daily development, operations, and maintenance decisions.

6.2 OPTIMIZATION APPROACHES

In 2011, *New Technology Magazine* (Cope, 2011), David Millar, senior vice-president, reservoir optimization, SPT Group stated, "Currently optimization is applied to perhaps 10% of oil and gas modeling scenarios. I foresee that within a decade the majority of modeling cases will use optimization." With the aggressive expansion of IAM projects and DOF initiatives, process optimization techniques are rapidly gaining ground in the oil and gas industry. Over the past several years, many new software tools and optimization methods have emerged and have been deployed in assets worldwide. For example, the implementation of top-down reservoir models (TDRM) (Williams et al., 2004) has been proven to add 20% to the net present value (NPV) and has proven significantly more efficient than the manual history-matching process, when applied to the optimization of the carbonate reservoir in the North Sea. While manual history matching gave very poor results due to >80 variables/parameters, the TDRM approach produced an acceptable match in less than a month and showed 50 million to 150 million more barrels of oil than the manual approach (Cope, 2011). Moreover, the use of a multipurpose environment for parallel optimization (MEPO) (Schulze-Riegert et al., 2001) has helped operators analyze—in half a day—hundreds of operating scenarios and proposed the NPV range solutions with the worst case of $13 million and the best case of $27 million (Cope, 2011). In addition, numerous oil and gas optimization studies have recently been published in the literature and deployed for these purposes:

- Dynamic optimization of waterflooding with smart wells using optimal control theory, a gradient-based optimization technique (Brouwer and Jansen, 2002; Brouwer et al., 2004) and adjoint-based optimal control (Sarma et al., 2006, 2008; Suwartadi et al., 2011).
- Waterflood performance management and optimization using data-driven predictive analytics from capacitance resistance models (CRM) for rapid characterization of inter-well connectivity and continuous update of injection rates to maximize oil production and recovery (Kansao et al., 2017).
- Multiobjective optimization with applications to model calibration and uncertainty quantification (Schulze-Riegert et al., 2007).
- Closed-loop production optimization and management using robust, constrained optimization of short- and long-term NPV (Wang et al., 2007; Chen et al., 2012) and ensemble-based optimization, using data assimilation techniques (Chen et al., 2009).
- Real-time optimization and proactive control of waterflood performance in intelligent wells, equipped by wellbore pressure and temperature sensors and inflow control valves (ICVs) (Temizel et al., 2015) and real-time reservoir management using multiscale adaptive optimization and control (Saputelli et al., 2006).
- Top-down intelligent reservoir modeling (TDIRM) (Gomez et al., 2009), which integrates traditional reservoir engineering analysis with artificial intelligence and data mining [artificial neural networks (ANNs), fuzzy sets] to predict reservoir performance and optimize development strategies.

Recent developments of optimization methods can be mainly divided into deterministic and heuristic approaches. Deterministic approaches leverage analytical properties of the problem and generate a sequence of points that converge to a global optimal solution. Heuristic (often referred to as probabilistic or stochastic) approaches are perceived to be more flexible and efficient than deterministic methods; however, they tend to be more computationally demanding most often since the probability of finding the global solution decreases when the problem size increases.

Prevalent deterministic optimization approaches are linear programming, mixed-integer linear programming (MILP), nonlinear programming, and mixed-integer nonlinear programming (MINLP). The application of linear programming in the oil and gas industry is to a large extent a matter of the past with publications dating back to the late 1990s (Eeg and Herring, 1997), because they are most effective in solving (rather scarce) linear optimization problems. One of the most comprehensive IAM planning solutions developed using MILP was by Iyer et al. (1998), which proposes planning and scheduling of investment and operation in offshore oil-field facilities by the rigorous incorporation of nonlinear reservoir performance, surface pressure constraints, and drilling rig resource constraints. Iyer et al. (1998) introduce a tractable MILP model with several thousand binary variables and sequential model decomposition strategy by the (dis)aggregation of time periods and wells.

However, nonlinear programming and particularly MINLP can provide general tools for solving optimization problems to obtain a global or an approximately global optimum and are still actively pursued, for example, in well-spacing optimization by maximizing NPV (John and Onjekonwu, 2010) and for generalized field development optimization in terms of well-drilling locations and corresponding (time-varying) controls (Isebor et al., 2014).

The rest of this section briefly reviews the mainstream optimization approaches in exploration and production (E&P). For a comprehensive overview of different optimization methodologies for decision making in intelligent DOF, see Echeverria Ciaurri et al. (2012). For a technology survey of real-time optimization of offshore oil and gas production systems, see Bieker et al. (2006). Temizel et al. (2014) outline the most prevalent advantages and drawbacks of optimization techniques in real-time production optimization of intelligent fields. Finally, for a comprehensive general review of optimization techniques we recommend the encyclopedia of optimization (Floudas and Pardalos, 2009) as an excellent resource.

6.2.1 Single- vs. Multiobjective Optimization

A mathematical optimization problem essentially combines three components:
- objective or cost function,
- optimization constraints, and
- control variables.

The objective of optimization is to determine a feasible combination of optimization (control) variables within the boundaries of defined constraints that maximizes or minimizes the objective or cost function of choice.

Based on the nature of the search for an optimal value of the objective function, the optimization problems can be classified as a single- or a multiobjective optimization problem. An example of a single-objective optimization is finding an extrema, a minimum, or maximum, of a nonlinear convex problem, such as a quadratic function. A common oil and gas optimization problem is (dynamic) model calibration or history matching, which seeks a least-square fit of reservoir simulation response to the observed or measured data. The misfit objective function Q is represented as (Ferraro and Verga, 2009):

$$Q = \sum_{i=1}^{n} R_i^2 \tag{6.1}$$

with $R_i = w_i(X_m - X_o)_i$ defined as a residual, where X_m, X_o, and w_i correspond to the model data (reservoir simulation response), observed (measured) data (e.g., pressures, fluid rates, gas-oil ratio) and the weighting factor, respectively.

The optimization problem can be approached as a single-objective optimization in which an aggregate of all the quantities to be matched are grouped into a single, joint objective function, or as a multiobjective optimization approach, which usually considers two or more different objectives, addressed separately during the optimization process. Mathematically, the single-objective optimization is defined as (Hutahaean et al., 2015)

$$\begin{aligned} & \text{minimize} f(\boldsymbol{x}) \\ & \text{subject to } h_k^l \leq x_k \leq h_k^u \\ & x = \{x_1, x_2, \ldots, x_k, \ldots, x_N\} \end{aligned} \tag{6.2}$$

where $x = \{x_1, x_2, \ldots, x_k, \ldots, x_N\}$ is the vector of the N variables in the parameterization and h_k^l and h_k^u, respectively, correspond to the lower and

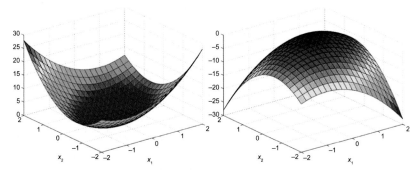

Fig. 6.2 Examples of objective functions in single-objective optimization indicating minimum (left) and maximum (right) as extrema.

upper boundaries of each variable. A geometric representation of objective functions for single-objective optimization problem is given in Fig. 6.2.

In production optimization problems, the NPV is generally used as an objective or cost function Q, subject to maximization. Following Wang et al. (2009) and Suwartadi et al. (2011) the NPV is mathematically formulated as

$$Q = \sum_{n=1}^{N} \left[\sum_{j=1}^{N_{prod}} \left(\frac{r_o q_{o,j}^n - r_w q_{w,j}^n}{(1+b)^{t^n}} \right) - \sum_{l=1}^{N_{inj}} r_{w,inj} q_{inj,l}^n \right] \triangle t^n \qquad (6.3)$$

where N is total number of reservoir simulation time-steps, N_{prod} is the total number of producing wells, N_{inj} is the total number of injectors, r_o is the oil revenue (USD/STB), r_w is the water production cost (USD/STB), $r_{w,inj}$ is the water injection cost (USD/STB), $q_{o,j}^n$ and $q_{w,j}^n$ are average oil and water production rates of the jth producer (STB/D) over the nth time-step, respectively, $q_{inj,l}^n$ is the average injection rate of injector l (STB/D) over the nth time-step, b is the annual interest rate (%), t^n is the cumulative time up to the nth time-step (year), and $\triangle t^n$ is the time interval of the nth time-step (day). A detailed review and evaluation of different types of objective functions in production optimization and history matching workflows is given in Mata-Lima (2011).

Traditionally, the solution of the maximizing NPV in oil and gas production optimization has been through applying optimal control theory (Brouwer and Jansen, 2002). The literature mainly refers to two categories of algorithms used to solve this problem:

- *Gradient-based algorithms*, where the gradients are derived from the adjoint method (Brouwer et al., 2004; Sarma et al., 2006, 2008;

Alpak et al., 2015; Ramirez et al., 2017). In optimal control theory the dynamic system, which usually corresponds to a nonlinear operator such as finite-difference reservoir simulator, is introduced in the objective function along with optimization constraints with a set of Lagrangian multipliers, $\lambda(n)$. In Lagrangian notation, the modified NPV objective function (see Eq. 6.2) now becomes

$$\tilde{Q} = \sum_{n=1}^{N-1} Q(n) + \lambda(n+1)^{T} \mathbf{g}(n) \qquad (6.4)$$

where n corresponds to the time-step of the simulation performed with reservoir simulator $\mathbf{g}(n)$. The steepest descent (Brouwer et al., 2004) or steepest ascent (Wang et al., 2009) method can be used to update the estimates of the optimal control vector. For example, Sarma et al. (2006) demonstrate that the approach is quite efficient and can render a 70% increase in cumulative oil production in open-loop implementation and a 60% increase in closed-loop implementation. However, a disadvantage of the adjoint approach is that it requires explicit knowledge of the model equations as well as extensive programming to implement the equations.

- *Gradient-free algorithms*, where solving the optimization problem is independent of the model equations used and does not require implementation of adjoint equations. They can be beneficial when large-scale distributed computing resources are scarce or not available and when the optimization problem suggests using many starting points. Lorentzen et al. (2006) develop a gradient-free approach using ensemble-based statistics, more specifically ensemble Kalman filter (EnKF) to optimize the NPV and total cumulative oil production. In addition, Wang et al. (2002) use the partial enumeration method (PEM), a discrete nongradient-based method, and Isebor et al. (2014) use particle swarm optimization (PSO) with mesh-adaptive direct search (PSO-MADS) for NPV optimization. Echeverria Ciaurri et al. (2011) use derivative-free (i.e., noninvasive, blackbox) metaheuristic methods (e.g., PSO) for optimization of oilfield operations, specifically the well choke settings.

In the last decade, powered by the rapid evolution of high-performance distributed and parallel computing (HPC), the multiobjective optimization is being used more often in the oil and gas industry. The multioptimization problem generally consists of addressing two or more different, and usually conflicting, cost functions. An example of conflicting objective functions is an NPV of an asset, of which exploitation is considered as an "opportunity", and some measure of the risk, associated with that exploitation (Echeverria Ciaurri et al., 2012). Such optimization problems involving multiple conflicting objectives are often addressed by "aggregating the objectives into a scalar function and solving the resulting single-objective optimization problem" (Schulze-Riegert et al., 2007). In multiobjective optimization, the selection of weights corresponding to specific components of single objective function (see Eq. 6.1) is omitted by splitting the objective function into several components, which are optimized simultaneously. The objective function now takes the form $F(x) = \{f_1(x), f_2(x), ..., f_k(x), ..., f_M(x)\}$ and the optimization problem is defined as (Schulze-Riegert et al., 2007; Hutahaean et al., 2015)

$$
\begin{aligned}
&\text{minimize } F(\boldsymbol{x}) \\
&\text{subject to } h_k^l \le x_k \le h_k^u \\
&x = \{x_1, x_2, ..., x_k, ..., x_N\}
\end{aligned}
\tag{6.5}
$$

where $F(x) : \mathfrak{R}^N \to \mathfrak{R}^M, x = \{x_1, x_2, ..., x_k, ..., x_N\}$ is the vector of the N variables in the parameterization, M is the number of objectives, and h_k^l and h_k^u, respectively, correspond to the lower and upper boundaries of each variable.

In contrast to single-objective optimization, the task now becomes finding a set of optimal solutions, also referred to as the Pareto optimal set, usually represented as a Pareto front (Fig. 6.3). Because different objectives in multiobjective optimization are not comparable, the concept of Dominance and Pareto optimality applies (Hutahaean et al., 2015). Fig. 6.3 shows this concept, while Fig. 6.4 gives an example of the objective space behavior for several iterations of a history-matching workflow, which combines the joint misfit minimization of field watercut and field static pressure, using the multiobjective genetic algorithm (MOGA) (Kam et al., 2016; Ferraro and Verga, 2009) with the population of 40 model realizations and by parameterizing reservoir permeability and water saturation.

Despite the success the multiobjective algorithms have demonstrated in optimization problems, such as history matching of reservoir simulation models, their performance is known to reduce substantially when the number of objectives in multiobjective function exceeds three. Hutahaean et al. (2017) refer to such problems as many-objective problems (MaOP), where Pareto-based algorithms become significantly less effective in discriminating between solutions. This compromises the concepts of Dominance and Pareto optimality as well as the convergence of the search procedure. Hutahaean et al. (2017) identify various possible conflicts between the

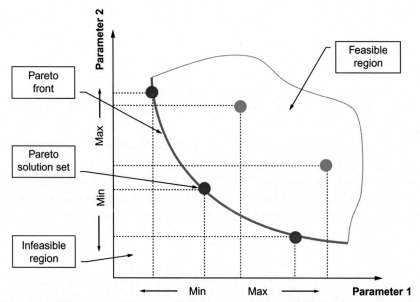

Fig. 6.3 Illustration of the concept of Dominance and Pareto optimality in objective space. *(Modified from Schulze-Riegert, R.W., Krosche, M., Fahimuddin, A., Ghedan, S.G., 2007. Multi-Objective Optimization with Application to Model Validation and Uncertainty Quantification. SPE-105313-MS. https://doi.org/10.2118/105313-MS.)*

objectives (direct, min–max, and nonparametric) and propose a nonparametric, conflict-based objective grouping to obtain faster and more robust history matches with better quality.

6.2.2 Local vs. Global Optimization

Specifically when solving multiobjective optimization problems, the selection of optimization technologies based on the nature of search for an optimal cost function value becomes highly relevant and sets the distinction between the *local* and *global* optimization. Note that the literature frequently prefers the term "global search" over "global optimization." The reason for this preference is that "…finding the global optimum in practical situations where the cost function is relatively time demanding and the number of optimization variables is larger than few tens is an extremely arduous (and virtually impossible in most cases) task. Hence, at most we can aspire is to search globally…" (Echeverria Ciaurri et al., 2012).

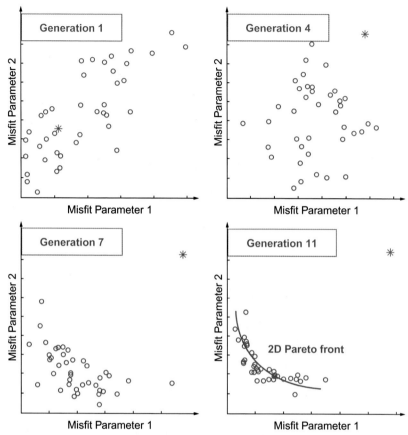

Fig. 6.4 Behavior of the two-component objective space in a history matching-workflow using the multiobjective genetic algorithm (MOGA) as a function of a number of minimization iterations.

The main attributes of local optimization methods can be summarized as follows:
- they generally seek a local solution and depend on derivatives of the cost function and constraints;
- the solution is not guaranteed to have the lowest objective (when minimization) or the highest objective (when maximization) among all the feasible points;
- they can be solved relatively easily, because they require the differentiability of the objective and constraints;
- they require initial guess (estimate); and
- they are frequently supported by a solid convergence theory.

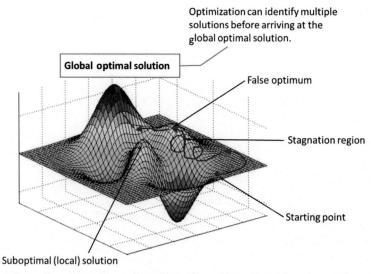

Fig. 6.5 Example of a complex multiobjective function for global search or optimization.

There are some cases (e.g., convex problems; see Fig. 6.2) where the local extrema found will in fact represent the global extrema; however, large-scale reservoir models with complex wells and facility networks usually render complex, multiobjective optimization problems. Fig. 6.5 illustrates an example of one such multiobjective cost function (such as the NPV of reservoir system).

As indicated, it combines several stagnation points, false optima, and suboptimal (local) solutions, all markedly different from the global objective, that is, maximized NPV. The complexity of multiobjective problems will drive engineers, as well as local optimization techniques, to stop the search once they have found a "plausible" solution. The global optimization (or search) on the other hand can identify multiple solutions to a range of engineering problems before reaching the global optimal solution. Attempts have been made to use the multistart methods combined with local optimization to generate multiple solutions with some degree of global search (Basu et al., 2016) or to deploy advanced proxy-based methods like Hamiltonian Markov chain Monte Carlo (McMC) (Mohamed et al., 2010a; Goodwin, 2015; Goodwin et al., 2017), surrogate reservoir models as smart proxies (Mohaghegh et al., 2015), and physics-based, data-driven models (Klie, 2015) to provide faster solutions to global optimization problems.

6.2.2.1 Stochastic or (Meta) Heuristic Optimization

The majority of global search/optimization methods in one way or another leverage stochastic/(meta)heuristic algorithms (Ombach, 2014) to explore large-scale optimization space, to reduce the probability of being trapped in the local solution that does not satisfy the global cost function, and to quantify the uncertainty of parameters in the sampled domain (Mohamed et al., 2010a). Many global search/optimization algorithms belong to the group of population-based methods described in great detail in Hajizadeh (2011) for improved history matching and uncertainty quantification in petroleum reservoirs. Echeverria Ciaurri et al. (2012) provide a condensed list of the main characteristics of population-based methods:

- Multiple points [i.e., solutions (see Fig. 6.4 for MOGA application)] are evaluated in every global search iteration.
- Unlike in pattern-search methods, these points are not clearly structured and can be, from iteration to iteration, rearranged with a much larger degree of flexibility.
- There are no theoretical results or empirical rules of thumb that recommend the population size for a given optimization problem, but it can be anticipated that the larger the size, the more globally the search space is explored. This is of course under the assumption that the space of optimization variables has been parameterized with sufficient degree of variability to allow a robust uncertainty quantification.
- Consistent with the above observation, if the population size is very small (compared to the number of optimization variables) the impact of these population-based methods will be locally confined.

A variety of stochastic/(meta)heuristic global search methods have recently emerged in applications to reservoir characterization (Compan et al., 2016) or production optimization, which are summarized in Table 6.1. The table lists a selected number of (meta)heuristic optimization methods with associated engineering application and relevant references.

6.2.3 Optimization Under Uncertainty

The process of optimizing reservoir performance under the assumption that all the system variables are deterministically known is relatively straightforward. However, with the presence of physical and financial uncertainties the problem is elevated to optimization with risk-managed, decision-making focus. According to McVay and Dossary (2014), the value of "reliably quantifying uncertainty is reducing or eliminating both expected disappointment (ED), when realizing an NPV is substantially less than estimated NPV, and

Table 6.1 Selected (Meta)heuristic Optimization Methods With Main Applications

Metaheuristic Technique		Optimization Application	Reference
Evolutionary algorithms (EA)	Single pareto evolutionary algorithms (SPEA)	Computer-assisted history matching (AHM)	Ferraro and Verga (2009)
	Genetic algorithms (GA)	Field development and production strategy optimization	Sambo et al. (2016) and Gomez et al. (2009)
	Evolution strategies (ES)	Well-placement optimization	Bouzarkouna et al. (2013)
	Multiobjective genetic algorithms (MOGA)	AHM Well-control strategy	Kam et al. (2016) and Fu and Wen (2017)
	Multi-island genetic algorithms (MIGA)	Reactive and proactive control for smart wells	Carvajal et al. (2014)
	Hybrid Genetic Algorithms (HGA)[a]	Production strategy optimization	Salam et al. (2015)
	Support vector machine (SVM) optimized by GA	Injection allocation optimization	Li et al. (2016)
Particle swarm optimization (PSO)		AHM	Arnold et al. (2016) and Mohamed et al. (2010b)
		Drilling optimization	Self et al. (2016)
		Well-placement and well-pattern optimization	Jesmani et al. (2015) and Onwunalu and Durlofsky (2011)
		Closed-loop field development optimization[b]	Shirangi and Durlofsky (2015)
Ant colony optimization (ACO)		AHM	Hajizadeh et al. (2009)
Simulated annealing (SA)		Well-scheduling and placement optimization	Beckner and Song (1995)

Table 6.1 Selected (Meta)heuristic Optimization Methods With Main Applications—cont'd

Metaheuristic Technique	Optimization Application	Reference
Markov chain Monte Carlo (McMC)	AHM	Schulze-Riegert et al. (2016), Maucec et al. (2007, 2011, 2013a,b), and Olalotiti-Lawal and Datta-Gupta (2015)[c]
	Reservoir description and forecasting	Li and Reynolds (2017)
	Prediction uncertainty quantification	Fillacier et al. (2014) and Goodwin et al. (2017)[d]
Differential evolution (DE)	AHM	Hajizadeh et al. (2010) and Olalotiti-Lawal and Datta-Gupta (2015)[c]
Tabu search (TS) and scatter search (SS)	Multiple-field scheduling optimization	Cullick et al. (2003)
	Project portfolio optimization	April et al. (2003)
	Well-placement optimization	Cullick et al. (2006)
	AHM	Yang et al. (2007)
	Artificial lift optimization	Vasquez et al. (2001)[e]

[a]HGA: hybrid technique of GA and ANNs.
[b]PSO-MADS: hybrid technique of PSO and mesh adaptive direct search (MADS).
[c]DEMC: hybrid technique of differential evolution (DE) and McMC.
[d]HMcMC: Hamiltonian McMC.
[e]GATS: hybrid technique of GA and Tabu search (TS).

expected decision error (EDE), through selecting the wrong projects". McVay and Dossary (2014) present a new framework for assessing the impact of overconfidence and directional bias on portfolio or asset performance. They further report that for moderate amounts of overconfidence and optimism, the ED amounted to 30%–35% of NPV for analyzed portfolios and optimization cases, which can profoundly affect the asset performance. In even broader context, Allen (2017) describes handling risk and uncertainty in portfolio and/or asset production forecasting. He builds on portfolio optimization under uncertainty and introduces sequencing of uncertainty and aggregation of risk as fundamental components in an asset's production vulnerability and associated risk management.

The many model and asset parameters of IAMod and IAM workflows are uncertain. This uncertainty introduces a new level of complexity in the optimization process, because optimization functions and/or constraints are no longer deterministic functions but probabilistic/stochastic distributions. However, if statistical principles are considered, the optimization frameworks as discussed above can be modified to incorporate stochastic functions. In reservoir characterization, an example of uncertainty quantification may represent a set (ensemble) of model realizations, each of them honoring a set of historic data. In retrospect, a stochastic production optimization problem may represent maximization of expected NPV over all available realizations. Note that the statistical nature of such a problem will render the mean (expected) NPV value with associated confidence intervals; however, the optimization will require many reservoir flow simulations and may be prohibitively time consuming. Echeverria Ciaurri et al. (2012) propose the approach of retrospective optimization (RO), which replaces a stochastic optimization problem by a sequence of optimization problems where constraining statistics are approximated with gradually increasing levels of quality.

An alternative approach is the use of stochastic programming (SP) (Nemirovski et al., 2009) where the optimization problem with objective function $F_0(x)$ formulates as follows:

$$\begin{aligned} \text{minimize } & \boldsymbol{F}_0(x) = \mathbf{E}f_0(x, \omega) \\ \text{subject to } & \boldsymbol{F}_i(x) = \mathbf{E}f_i(x, \omega) \leq 0, \ i = 1, \dots, m \end{aligned} \tag{6.6}$$

where $\mathbf{E}$ represents the expected value operator on objective and constrain functions $f_i(x, \omega)$, which depend on x and ω, optimization and random variables, respectively. The value of ω is not known, but its distribution is and the goal is to select x so that constraints are satisfied on average or with high probability and the objective is minimized on average or with high probability. The stochastic constraint $\mathbf{E} f(x) < 0$ is classified as a standard quadratic inequality.

It is interesting to note that neither RO nor SP are markedly represented in the area of oil and gas production optimization problems. However, the E&P industry has been rapidly adopting a complementary ensemble-based approach to assisted history matching (AHM) with uncertainty, using for example Bayesian inversion techniques such as ensemble Kalman filter (EnKF) (Evensen, 1994), ensemble smoother with multiple data assimilation (ES-MDA) (Emerick and Reynolds, 2013), or sequential, Markov-chain

Monte Carlo (McMC) (Maucec et al., 2007, 2011, 2013a,b). These techniques are outlined in more detail in later sections.

In 2001, Begg et al. (2001) outlined the need for a holistic, integrated approach to assess and manage the impact of uncertainties for the optimization of oil and gas assets and investment decision-making. They proposed the concept of a stochastic integrated asset model (SIAM) embedded in a decision-support system. While the SIAM ties to a classical domain models via a simulation engine (reservoir simulator) as a key component to quantify uncertainties and their nonlinear behavior, the new components of the workflow-like scenario analysis, real options thinking/valuation, value of flexibility, and decision optimization are introduced. In following years, the E&P industry has adopted and applied a variety of approaches to asset production optimization with uncertainty:

- Bailey et al. (2004) introduced the workflow for NPV optimization under uncertainty through the utility function $F_\lambda = \mu - \lambda\sigma$, where μ and σ are the mean and standard deviation of the collection of N individual appropriately sampled NPV realizations, and λ represents the risk-aversion factor. The utility function F_λ implies a maximization of NPV under the constraint of the variable-importance minimization of its standard deviation, depending on the user's own risk preference.

- Cullick et al. (2004) designed an optimization system that consists of three workflows: an outer optimization workflow (outer loop), an inner scenario and uncertainty-management-simulation workflow (inner loop), and a dispatcher for distributed computing. The optimizer uses (meta)heuristic search methods and is validated with three problems of asset optimization: intuitive solution, nonintuitive solution without uncertainty, and solution with URM.

- Sarma et al. (2005) introduced a closed-loop production optimization under uncertainty with polynomial chaos expansion for an efficient uncertainty propagation and validated the workflow in real-time optimization of NPV for the reservoir-under-waterflood regime, production constraints, and uncertain subsurface characterization.

- Liu and Reynolds (2015) use multiobjective optimization [weighted sum (WS) and normal boundary intersection (NBI)] for jointly maximizing the expectation and minimizing the risk by solving a max-min problem and applying it to a well-placement optimization and optimal well control. Here, the concept of minimizing the risk simply refers to "maximizing the minimum value life-cycle NPV".

- Shirangi and Durlofsky (2015) developed and apply a methodology for closed-loop field development (CLFD) under geological uncertainty. A multilevel hybrid optimization, combining metaheuristic techniques PSO with mesh adaptive direct search (MADS), is validated for simultaneous and sequential field development with maximizing NPV as the objective function. They further propose enhancements in the form of bi-objective optimization with minimizing the risk objective while maximizing the expected performance.
- Arnold et al. (2016) propose a comprehensive, field full lifetime workflow for uncertainty propagation and rigorous optimization of decision-making under uncertainty, applied to assets with naturally fractured reservoirs. They use the design of experiments (DoE) to perform sensitivity analyses and distance-based multidimensional scaling (MDS) to identify and dynamically rank (Maucec et al., 2011) the model candidates, multipoint statistics (MPS) to generate updates of the discrete fracture network (DFN) models in the field appraisal phase, the multiobjective PSO (MOPSO) for history matching and in development, history matching and Naïve Bayes (NA-Bayes) for unbiased estimation of uncertainty for forecasting and reservoir management phases of the field lifetime. This novel attempt at technology integration results in <5000 reservoir simulation iterations to attain a robust set of decisions for a complete field lifetime.

6.3 ADVANCED MODEL CALIBRATION WITH ASSISTED HISTORY MATCHING

The calibration of reservoir simulation models to dynamic field production data, commonly referred to as dynamic data integration or history matching, is perceived as one of the most time-consuming engineering processes in reservoir validation. In a *New Technology Magazine* article (Cope, 2011), Maucec stated, "History matching must be considered as a bridge between the reservoir modeling and reservoir simulation." Traditionally, reservoir models were manually reconciled with production data, using good engineering judgment and following workflows based on many years of experience.

The main disadvantage of the manual history-matching process is that it disengages the reservoir simulation model from the geological model and, many times, fails in adequate quantification of reservoir uncertainty. As a result, manual history matching frequently leads to unrealistic, nongeological and nonphysical features in the reservoir model. Moreover, the lessons learned are not properly applied to create a realistic reservoir model, and the perceived "history-matched" models are of low predictive value.

As reservoirs and assets mature and data acquisition and processing methods evolve and become more sophisticated, the acquired reservoir data grow substantially in terms of quantity and complexity. Particularly with the expansion of DOF projects that use automated and data-driven IAMod and IAM workflows, the need to solve large-scale, high-resolution modeling problems, quantify the inherent model uncertainty for more reliable prediction, and optimize their performance are becoming prevalent in the E&P industry. The challenges resulting from integrating multiple scales of data with uncertainties in physical parameters and processes make imperative the use of efficient model parameterization, advanced inversion and optimization algorithms, with utilization rapidly evolving HPC architectures.

Within the last three decades, the oil industry has gained traction in developing and implementing stochastic, population-based algorithms in reservoir characterization and simulation workflows. The applications of simulated annealing (SA), an algorithm that was originally developed for solving combinatorial optimization problems, first emerged in the oil and gas industry in the early 1990s in areas from stochastic reservoir modeling to optimization of well-scheduling and -placement (Deutsch and Journel, 1994; Ouenes et al., 1994) and have endured through the introduction of advanced SA algorithms, such as very fast simulated annealing (VFSA), with recent expansion of unconventional exploration (Sui et al., 2014).

Another important advance in oil and gas stochastic modeling was the introduction of techniques for the design of experiments (DoE), which was originally developed in agriculture in the late 1920s (Salsburg, 2001). DoE modeling has been primarily used for the rapid quantification of uncertainty using proxy models with response surface analysis (RSA) and various forms of designs (e.g., latin hypercube, Box-Behnken, etc.) in AHM (Cullick et al., 2004; Alpak et al., 2013), sensitivity analyses (Fillacier et al., 2014), and risk evaluation (Sazonov et al., 2015). In the early 2000s the E&P industry started to see an expansion of ensemble-based, Bayesian inference and model inversion, using, for example, the evolutionary algorithms (Schulze-Riegert and Ghedan, 2007), the ensemble Kalman filter (EnKF) (Evensen, 2009), and recently a complementary data assimilation approach, the ensemble smoother (ES) and multiple data assimilation (MDA) by Emerick and Reynolds (2012, 2013) and Maucec et al. (2016, 2017) and the EnKF coupled with the streamline sensitivity-based covariance localization (Arroyo et al., 2008). Several techniques and workflows have been successfully developed in this area, such as methods based on sequential (or random-walk) McMC, originally developed in the areas of statistical physics (Neal, 1993).

The McMC method is arguably the most rigorous statistical approach to sample from the stationary Bayesian distribution; however, when deployed in direct simulation, it imposes a high computational cost. To improve the performance of the McMC method, several enhancements were proposed, based on a two-step proposal of jointly sampling the model and data variables (Oliver, 1996), by constraining the proxy models using the streamline sensitivities (Efendiev et al., 2005; Ma et al., 2006; Maucec et al., 2007, 2013a,b) or by coupling with adjoint methods (Schulze-Riegert et al., 2016). These techniques enhance the sampling efficiency of the McMC method and make it applicable for the inversion of large-scale reservoir models without sacrificing scientific rigor.

Recently, Goodwin (2015) proposed an alternative to random-walk McMC, namely Hamiltonian McMC techniques which progress rapidly through the sampled space but require derivatives of likelihood that can be efficiently implemented with proxy models. In parallel, the development of AHM tools and approaches has also evolved toward "smart" proxy models in the form of surrogate reservoir models (SRM) (Mohaghegh et al., 2015) and increasingly popular (meta)heuristic methods, such as PSO (Mohamed et al., 2010a) and differential evolution (Hajizadeh et al., 2010). Moreover, developments in the area of AHM are also leading toward joint inversion of the production and time-lapse seismic data, where the attributes of four-dimensional (4D) seismic inversion (e.g., water saturation) can provide spatially rich information on the fluid flow dynamics within subsurface reservoirs (van Essen et al., 2012; Jin et al., 2012). While the resulting reservoir model updates exhibit a considerable improvement in matching the saturation distribution in the field, the potential drawback is the dependence on the inversion data from 4D seismic surveys, which are difficult and expensive to obtain.

This section continues with a review of modern model calibration and inversion techniques and then describes the E&P industry's prevalent model parameterization, AHM, and finally outlines a few applications of IAM wokflows. For further reading, Schulze-Riegert and Ghedan (2007), Oliver and Chen (2011), and Rwenchungura et al. (2011), among others, provide comprehensive overviews of recent advancements made in the area of reservoir AHM, while Hajizadeh (2011) gives a very comprehensive review of the evolution of history matching over the last 50 years.

6.3.1 Model Parameterization and Dimensionality Reduction

Dynamic calibration of reservoir models is an ill-posed and potentially unstable inverse problem, where many probable solutions can satisfy the posed objective function. This approach often requires a re-definition of subsurface spatial properties into parameter groups that provide a tool for problem regularization and make the inversion problem computationally tractable. Such techniques are commonly referred to as "parameterization techniques" and in their applications to history matching and model calibration their goal is primarily to replace the original set of unknown spatially discretized reservoir properties with a reduced (lower-dimensional) number of parameters while retaining minimum possible loss of information density in representing the most representative features of the reservoir model.

An example is a parameterization of the depositional system and underlying facies distribution to strategically group the most dominant flow units while retaining the inherent spatial heterogeneity and continuity that drive the reservoir connectivity. Bhark et al. (2011) provide an in-depth review of the most prevalent model parameterization techniques, while Hutahaean et al. (2015) and Al-Shamma et al. (2015) study their impact on the objective choices with implications to simulation model history matching.

Zonation (and its adaptive variants) has traditionally been used in petroleum applications for adjusting reservoir model static properties like porosity, permeability, and transmissibility. Herewith, however, we briefly present the subspace and low-rank approximation methods that provide the basis for multiscale parameterization based on linear transformation for applications in AHM and optimization. In brief, the methods of linear transformation map the spatial parameters (i.e., reservoir parameters, like porosity or permeability, subject to parameterization) from the "parameter domain" to the transformed, low-rank (i.e., low dimension) "feature domain", where parameter estimation and model updating can be performed in a more efficient manner. Using an orthogonal transform, the discrete spatial variable $\mathbf{u}$ is mapped to the transform domain as (Bhark et al., 2011).

$$\mathbf{v} = \mathbf{\Phi}^T \mathbf{u} \Leftrightarrow \mathbf{u} = \mathbf{\Phi}\mathbf{v} \qquad (6.7)$$

where vector **u** has a dimension of $m \times 1$ (m = discretization of the reservoir parameter, that is, the total number of representative grid cells) and the column vector **v** is the n_t-length spectrum of transform coefficients. The n_t columns of the transform basis **Φ** represent the discrete basis functions with length of m. The main objective of parameterization is to reduce the parameter dimension, that is, the dimension of vector **v**, with a compact/truncated representation of **Φ** that contains only a few basis functions that are still able to capture relevant model spatial information. Schematically, the parameterization by linear transformation mapping is presented in Fig. 6.6, while Table 6.2 lists the prevalent subspace model parameterization methods

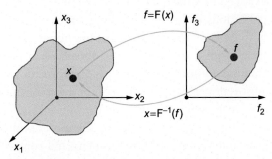

Fig. 6.6 Schematic representation of mapping from the parameter space to the feature domain.

Table 6.2 Selected (Meta)heuristic Optimization Methods With Main Applications Parameterization

Technique	Reference
Singular value decomposition (SVD)	Yanai et al. (2011)
Karhunen-Loeve transform (KLT)	Newman (1996) and Jafarpour and McLaughlin (2007)
Fourier-space filter expansion	Maucec et al. (2007)
Principal component analysis (PCA)	Honorio et al. (2015), Kang et al. (2015), and Chen et al. (2014)
Discrete cosine transform (DCT)	Jafarpour and McLaughlin (2007, 2009) and Maucec et al. (2011, 2013a,b)
Grid-connectivity transform (GCT)	Bhark et al. (2011), Kang et al. (2014), and Kam et al. (2016)
Multidimensional scaling (MDS)	Scheidt and Caers (2009), Maucec et al. (2011, 2013a,b), and Arnold et al. (2016)

Note: Parameterization techniques SVD, KLT, and PCA are occasionally commonly referred to as proper orthogonal decomposition (POD) techniques.

applied to petroleum studies. The reader is encouraged to consider listed references for further mathematical detail.

6.3.2 Bayesian Inference and Updating

History matching is a highly nonlinear and ill-posed inverse problem. This means that depending on the prior information, a set of nonunique solutions can be obtained that honor both the prior constraints and conditioned data with associated uncertainty. To assess the uncertainty in estimated reservoir parameters, one needs to sample the parameters of the posterior distribution. The Bayesian method provides a very efficient framework to perform this operation.

Using the Bayes' formula, the posterior distribution (i.e., the probability of occurrence of model parameter m (simulated) given the data d [observed or measured)] can be represented as proportional to the product of the likelihood function and a prior probability distribution of the reservoir model:

$$p_{m|d}(\mathbf{m}|\,\mathbf{d}) = \frac{p_{d|m}(\mathbf{d}|\,\mathbf{m})p_m(\mathbf{m})}{p_d(\mathbf{d})} \tag{6.8}$$

where $p_{m|d}(\mathbf{m}|\mathbf{d})$, $p_{d|m}(\mathbf{d}|\mathbf{m})$, and $p_m(\mathbf{m})$ represent the posterior, likelihood, and prior probability distribution, respectively. The normalization factor $p_d(\mathbf{d})$ represents the probability associated with the data. It is independent of the model parameters and is usually treated as a constant.

The Bayesian formulation of objective function combines observed data with the prior geological information. It is assumed that the "prior model" parameters follow a multi-Gaussian probability density function (pdf) with a prior covariance matrix of reservoir model parameters, $\mathbf{C_M}$. Therefore, its pdf, $p_m(\mathbf{m})$, is given by

$$p_m(\mathbf{m}) = \frac{1}{(2\pi)^{M/2}|\mathbf{C_M}|^{1/2}} \exp\left[-\frac{1}{2}(\mathbf{m}-\mathbf{m}^0)^{\mathrm{T}}\mathbf{C_M^{-1}}(\mathbf{m}-\mathbf{m}^0)\right] \tag{6.9}$$

This distribution is centered around the prior mean $\mathbf{m}^0$. The "likelihood function," defined as the conditional probability density function of the data given the parameters $p_{d|m}(\mathbf{d}|\mathbf{m})$ is calculated using

$$p_{d|m}(\mathbf{d}|\,\mathbf{m}) = \frac{1}{(2\pi)^{N/2}|\mathbf{C_d}|^{1/2}} \exp\left[-\frac{1}{2}(\mathbf{d}-\mathbf{g}(\mathbf{m}))^{\mathrm{T}}\mathbf{C_d^{-1}}(\mathbf{d}-\mathbf{g}(\mathbf{m}))\right]$$

$$\tag{6.10}$$

where $\mathbf{C_D}$ is the data covariance matrix. The relationship between the data and the model parameters is expressed as a nonlinear function that maps the model parameters into the data space, $\mathbf{d} = \mathbf{g}[\mathbf{m}]$, where $\mathbf{d}$ is the data vector with N observations representing the output of the model, $\mathbf{m}$ is a vector of size M whose elements are the model parameters, and $\mathbf{g}$ is the forward model operator, a function that relates the model parameters to the output. For history-matching problems, $\mathbf{g}$ represents the reservoir simulator.

Using Bayes' inference (Eq. 6.8), the posterior pdf can be defined as follows:

$$p_{m|d}(\mathbf{m}|\,\mathbf{d}) \propto \exp\left[-O(\mathbf{m})\right] \qquad (6.11)$$

where $O(m)$ represents the Bayesian OF in multi-Gaussian notation:

$$O(\mathbf{m}) = \frac{1}{2}(\mathbf{d}-\mathbf{g}(\mathbf{m}))^{\mathbf{T}}\mathbf{C_d^{-1}}(\mathbf{d}-\mathbf{g}(\mathbf{m})) + \frac{1}{2}(\mathbf{m}-\mathbf{m}^0)^{\mathbf{T}}\mathbf{C_M^{-1}}(\mathbf{m}-\mathbf{m}^0)$$

$$(6.12)$$

The first right-hand term of Eq. 6.12 represents the data misfit term (mismatch between observed data and simulated response), while the second right-hand term corresponds to a regularization term, usually represented by the prior/known geomodel. The result of minimizing the $O(m)$ (Eq. 6.12) is called the Maximum-A-Posteriori (MAP) estimate because it represents the most likely posterior model. The set of parameters m that minimizes $O(m)$ is the most probable estimate.

The goal of the Bayesian approach is to derive a statistical distribution for the model parameters via posterior distribution, constrained through the prior distribution. Because the reservoir history-matching problem is an inverse problem, its solution renders multiple plausible models (i.e., multiple realizations), and the consequence of nonlinearity is that one must resort to an *iterative* solution. The MAP estimate is often insufficient because it does not provide the uncertainty quantification in the posterior model. As a solution, Kitanidis (1995) and Oliver et al. (1996) introduced the randomized maximum likelihood (RML) method which provides a theoretically rigorous approach to sample from the posterior distribution but holds only for linear Gaussian problems, which are seldom the case for reservoir simulation model inversion. Several alternative approaches for a rigorous sampling from posterior distribution for nonlinear problems have been proposed, such as traditional Markov chain Monte Carlo (McMC) (Neal, 1993), its multistage implementation with enhancements of convergence efficiency (Efendiev et al., 2005), and the variant using sampling from the streamline-sensitivity constrained proxy model (Maucec et al., 2007, 2011, 2013a,b; Ma et al., 2006). The multistage McMC approach still satisfies the necessary detailed-balance condition to sample from the equilibrium (i.e., stationary or posterior) distribution (Maucec et al., 2007) and significantly faster convergence, which can be monitored via, for example, maximum entropy condition (Maucec et al., 2007; Bratvold et al., 2010) or by using multivariate potential scale reduction factor (MPSRF) (Li and Reynolds, 2017).

The McMC algorithm is designed to rigorously sample from the posterior distribution; its drawback lies in the reliance on the specification of the prior model statistics and also the computational cost in exploring the posterior distribution when the number of parameters is large. When realistic field conditions are considered, the number of parameters of the prior model expands dramatically (i.e., $\sim 10^6$). Computation of the prior term of the objective function then becomes highly demanding and time consuming, particularly due to inversion of the prior covariance matrix C_M. To maintain acceptable computation effort, the AHM algorithms have to resort to model parameterization techniques as outlined in Table 6.2. For example, Maucec et al. (2007) introduce the approach where parameterization and model reduction on a covariance matrix C_M is performed using SVD, and new model realizations are generated in the wave-number domain by a simple convolution of zero-mean independently distributed entries (white noise) with an appropriate Fourier linear filter. Jafarpour and McLaughlin (2007) introduce the application of the discrete cosine transform (DCT), an industry standard for image compression (e.g., the JPEG format) in reservoir model inversion and history matching, while Maucec and Cullick (2015) and Maucec (2016) develop a DCT-based approach for rapid generation of model updates in wave-number domain with applications to AHM (Maucec et al., 2011, 2013a,b).

6.3.3 Data Assimilation

One of the most popular data assimilation approaches in petroleum and groundwater applications is Ensemble Kalman Filter (EnKF), introduced in the early 1990s (Evensen, 1994) for forecasting error statistics. The EnKF was first deployed for solving history match reservoir simulation problems by Nævdal et al. (2003). It approaches the AHM optimization (minimization) problem by solving the Bayesian form of the objective function $O(m)$ in the multi-Gaussian notation (see Eq. 6.12) using the iterative Gauss-Newton method (Le et al. 2015):

$$\mathbf{m_{i+1}} = \mathbf{m_i} - \beta_i \left\{ \left(\mathbf{m^i} - \mathbf{m_{pr}} \right) + \left(\mathbf{C_M G}_i^T \right) \left[\mathbf{G}_i \mathbf{C_M G}_i^T + \mathbf{C}_D \right]^{-1} \right.$$

$$\left. \left[\mathbf{g}(\mathbf{m^i}) - \mathbf{d_{obs}} - \mathbf{G}_i(\mathbf{m^i} - \mathbf{m_{pr}}) \right] \right\} \tag{6.13}$$

where i is the iteration index, β_i is the iteration step-size in the search direction, and $\mathbf{G}_i$ is the sensitivity matrix at m_i. The term $\mathbf{C_M G^T [G C_M G^T + C}_D]^{-1}$ is usually referred to as the Kalman gain. The covariance terms in the ensemble update of Eq. (6.13) usually satisfy approximations $\widetilde{\mathbf{C}}_{MD} \approx \mathbf{C_M G}$ and $\widetilde{\mathbf{C}}_{DD} \approx \mathbf{G C_M G^T}$, where $\widetilde{\mathbf{C}}_{MD}$ corresponds to the cross-variance between the vector of model parameters m and the vector of predicted data d, while $\widetilde{\mathbf{C}}_{DD}$ is the auto-covariance matrix of the predicted data.

While EnKF is generally considered as a robust, efficient, and easy-to-implement tool for sequential data assimilation and uncertainty quantification, the main disadvantages of the method are the Gaussian approximation applied in the model update scheme and the suboptimal performance in terms of assimilation convergence when the relations between the (reservoir) model parameters and the data predicted by the forward estimator (e.g., reservoir simulator) are highly nonlinear. These issues may lead to a well-known problem of ensemble collapse, which is particularly evident for small ensemble sizes (Jafarpour and McLaughlin, 2009). Variants of ensemble design and update have been developed to alleviate these issues through more efficient handling of model constraints with Kernel-based EnKF (Sarma and Chen, 2011) as well as introduction of subspace EnKF and ES with Kernel PCA parameterization (Sarma and Chen, 2013). The ES also fits the category of ensemble-based data-assimilation methods but, in comparison to EnKF, which updates both model parameters and the states of the system, the ES is an alternative data assimilation method that computes the update of global reservoir model parameters in real time, by assimilating all data simultaneously. The ES was originally proposed by van Leeuwen and Evensen (1996) in an application with an ocean circulation model. Skjervheim et al. (2011) and Chen and Oliver, 2013 describe recent applications of ES for AHM of reservoir simulation models.

Emerick and Reynolds (2012, 2013) further propose the ES with the multiple data assimilation (ES-MDA) algorithm, which repeats the ES procedure several times on the same multiple observed data. Recently, the ES-MDA algorithms have been successfully deployed to integrated uncertainty analysis to render more robust field development decisions (Hegstad and Sætrom, 2014), reservoir modeling with integration of drill-stem (DST) data (Sætrom et al., 2016), AHM and rigorous uncertainty quantification in high-resolution model of a naturally fractured reservoir (Maucec et al., 2016), as well as for the AHM with uncertainty of a large-scale, dual porosity-dual permeability (DPDP) integrated reservoir model (IRM) (Maucec et al., 2017) powered by a massive parallel processing simulation platform.

6.3.4 Closed-Loop Model Updating

Closed-loop (reservoir or asset) model updating integrates the principles of (optimal) control theory and closed-loop optimization with (multiple) data assimilation into a workflow for reservoir optimization in terms of recovery or financial measures over the life of the reservoir using periodic, near real-time updates. A closed-loop controller operates with a so-called negative feedback loop that dynamically compares the system output with the reference point using sensor systems. The measured difference is channeled into a control device that dynamically applies the change to adjust the system input so it better matches its output.

In oil and gas exploration, the concepts of closed-loop model updating in a variety of workflows for reservoir management are presented in the open literature under different names, however, with quite similar objectives as stated above. Wang et al. (2007), Jansen et al. (2009), Chen et al. (2012), Barros et al. (2015), and Sampaio Pinto et al. (2015) introduce *closed-loop reservoir management (CLoReM)* (Fig. 6.7). Saputelli et al. (2006) and Sarma et al. (2005, 2006) refer to the similar principles of optimal control and model updating as *real-time reservoir management*. Dilib and Jackson (2013) describe it as a *closed-loop production optimization*. Oberwinkler and Stundner (2005) and Bieker et al. (2006) refer to the workflow as *real-time production optimization*, Saputelli et al. (2003) refer to it as *self-learning reservoir management*, while Hanea et al. (2015) present the development and implementation of the *fast model update (FMU)* workflow.

Frequently, the closed-loop model updating workflows as outlined above integrate the principles of performing direct modifications of the static geomodel parameters by the dynamic simulation modeling to perform the geologically consistent model update and optimization using the computer-

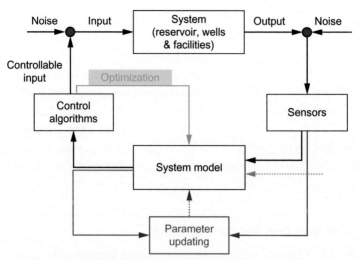

Fig. 6.7 Schematic representation of closed-loop reservoir management (CLoReM) workflow. *(With permission from Jansen, J.D., Douma, S.D., Brouwer, D.R., van den Hof, P.M.J., Bosgra, O.H., Heemink, A.W., 2009. Closed-Loop Reservoir Management. SPE-119098-MS. https://doi.org/10.2118/119098-MS.)*

assisted history matching (AHM) techniques under uncertainty. The objective of so-called "big-loop workflows" (Wiluweit et al., 2015; Kumar et al., 2017) is to generate reservoir simulation models that adequately quantify multiparameter static and dynamic uncertainties with associated risks and to render better predictive value for the field and asset development planning. Fig. 6.8 shows a schematic representation of a big-loop workflow.

6.4 OPTIMIZATION OF MODERN DOF ASSETS

This section outlines examples and applications of optimizing DOF assets using IAM workflows that integrate subsurface reservoir models and surface production network models. With rapidly expanding asset complexity driven by the size of the models and vast amounts of acquired and processed data, the major challenge of IAM still represents the coupling of the dynamic reservoir simulation model and the surface facilities into a single, integrated platform that allows the simultaneous simulation of the entire oil and gas system, "all in-one" and forecasts the asset's performance for the purpose of management.

The modern DOF system requires state-of-the art systems integration that can, for example, diagnose and address the operational challenges such

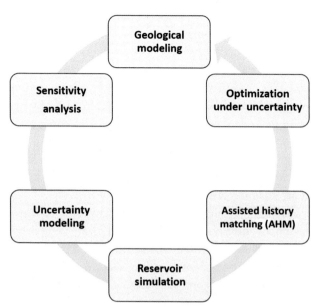

Fig. 6.8 Schematic representation of the big-loop conceptual workflow.

as monitor pressure distribution between subsurface and surface, identify system bottlenecks and backpressures, understand system constraints, manage mixing of fluids, and enable flow assurance (Ursini et al., 2010). In addition, the state-of-the-art IAM system should incorporate technologies to quantify effectively model uncertainties to provide an optimization and decision-making framework for managing the asset under uncertainty.

In practice, the modern IAM workflows can be classified in several categories that mainly differ in how the subsurface modeling applications communicate with surface network systems in terms of process automation, level of user interactivity, and the types of application coupling. These systems are applied to a variety of different field applications: oil, gas, stacked pay, multiple reservoirs, offshore and onshore fields, compositional and black oil, and different recovery processes:

• *Flow table coupling.* This workflow uses the tables of flowing bottom-hole pressure (BHP) as a function of flow rate for varying parameters such as gas-oil ratio (GOR), wellhead pressure (WHP), water cut (WCUT), lift gas, and pressure-volume-temperature (PVT) properties for interpolation by well models. The tables are set up to capture liquid rate changes with pump frequency or gas lift rate. The simulators use each well table during the entire life span of the well or completion event (e.g., depletion period and artificial lift completion). The simulator can run quickly but may sacrifice some accuracy related to production rate changes.

- *Static coupling.* This workflow uses a priori generation of reservoir performance tables comprising oil, gas, and water production rate forecasts over the desired time horizon for all production wells, by executing the reservoir model independently, and then providing the predicted rates as boundary conditions for the time-dependent execution of the surface models. Fig. 6.9 shows an example of a workflow with static coupling. This is the most commonly used coupling mode by operators today but is less accurate than dynamic coupling.
- *Dynamic coupling (loose coupling).* The reservoir and surface models are executed synchronously. At every time step, the reservoir model first predicts the production rates, which are then used by the surface models to generate the well boundaries for the execution of the reservoir model at the subsequent time step. The surface model predicts the flowing BHP (fBHP) (based on surface pressure) and the predicted value is imposed over the value to the simulator as a starting point for the convergence iteration. The simulator calculates a sandface pressure which meets the fBHP with an acceptable error. Fig. 6.10 shows an example of a workflow with dynamic (loose) coupling.
- *Tight iterative coupling.* Extends the dynamic coupling method described above to a more rigorous solution through an iterative approach. At every time step, the solution iterates between the pressure and flow boundaries at the sandface or at the wellhead, until convergence is achieved when the predicted pressure error between the two simulators

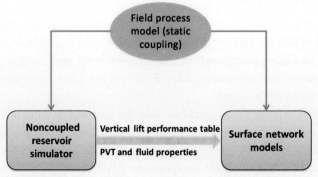

Fig. 6.9 An example of IAM workflow with static coupling of reservoir simulation and surface network models.

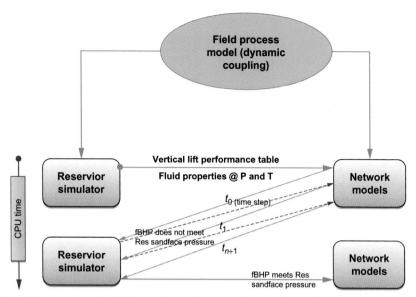

Fig. 6.10 An example of IAM workflow with dynamic (loose) coupling of a reservoir simulation model and an integrated surface network model.

remains below a specified threshold. This coupling mode requires extensive computational hardware and CPU time. In our experience, using static coupling, a black oil reservoir model with 10 wells (20 years of history) and using 8 CPU processors could run up to 10 min, dynamic coupling will run 100 min, and tight coupling could run >1000 min. Of course as technology improves the computational time decreases. A few examples of coupled simulators appear in the following: Fleming and Wang, 2017; Vanderheyden et al., 2016; and Khedr et al., 2012.

Fig. 6.11 compares production forecasts of oil, gas, and water rates generated by static and dynamic coupling of a reservoir simulator and a surface network system. As indicated, the production forecasts (e.g., oil, gas, and water rates) generated by dynamic coupling may be significantly different from the forecasts generated by the static coupling, based on the use of traditional rate forecast tables. The primary advantage of the dynamic coupling over static coupling is that the integrated reservoir simulation model is dynamically updated to reflect the constraints imposed by field operations over time.

The value of dynamically incorporating the effect of changing surface conditions on the reservoir model is that the solution generates a more realistic and accurate physical model of actual field constraints. As observed in Fig. 6.12, the simulation results using static coupling is slightly off from

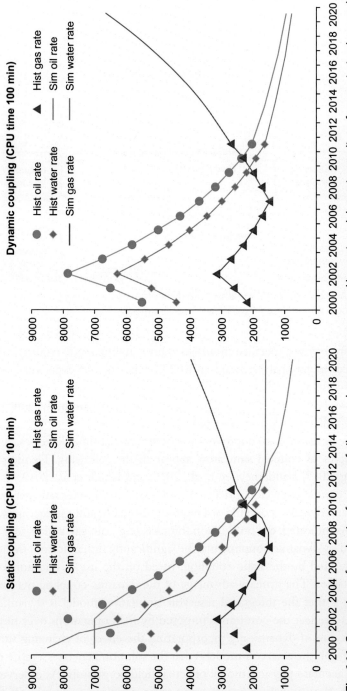

Fig. 6.11 Comparison of production forecasts of oil, gas, and water rates generated by static and dynamic coupling of reservoir simulator and surface network system.

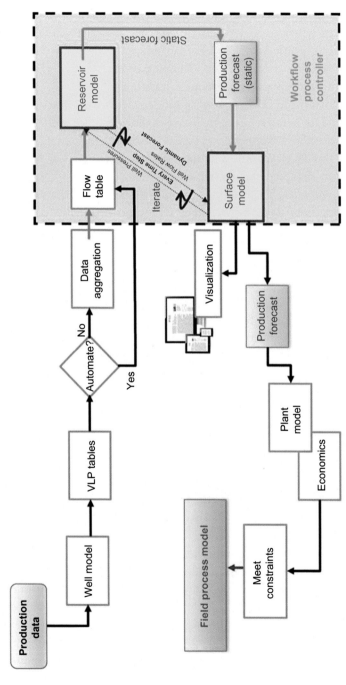

Fig. 6.12 An example of an IAM workflow with tight iterative coupling of a reservoir simulation model and surface network model.

historical data, particularly at the beginning of the field exploitation. The effect of surface constraints on the predicted production forecast is more pronounced when:
- the field is producing at or near capacity,
- field operating conditions are changing, or
- the simulation is over an extended duration.

The step forward in development of modern IAM workflows is the introduction of tight iterative coupling between the subsurface model and surface network workflow components; Fig. 6.12 shows an example. One novel aspect of this IAM workflow is the seamless integration of components, in the green-shaded rectangle, which are:
- *Data aggregation*. This module provides services for gathering and summarizing the data along with techniques for data quality assurance and quality control (QA/QC), in preparation for (statistical) analysis. Such techniques may include data cleaning and manipulation with removal of statistical outliers, imputation, and interpolation of missing data, etc.
- *Visualization*. This module provides data visualization in real time, which may include forecast, profiles, trends, and vertical flow performance (VFP) data. The visualization may have color maps of saturation, pressure, and composition, contours, and/or streamlines or flow vectors.
- *Workflow process controller*. This module provides interface and control platform for dynamic coupling between reservoir simulation models and surface network models. This workflow is designed to control well operations such as those using artificial lift pumps, automated surface chokes, downhole ICVs, etc. It also includes recovery process management such as water, gas, or chemical injection. Note that the production predictions are ultimately used in surface processing and economic models, which are not covered in detail here.

The value of deploying the IAM workflow with tight iterative coupling is significant; the workflow:
- Improves the integration efficiency by streamlining processes across multiple disciplines (reservoir, production, facilities, planning).
- Creates a more realistic production forecast prediction for planning and economics.
- Serves as an efficient, robust, flexible subsurface-to-surface integration building block for subsequent production optimization workflows; whatever event happens at surface affects the reservoir condition and vice versa.

- Facilitates development of "institutionalized" and "standardized" automated workflows.
- Serves as the integration platform where models can be executed and saved to a database for later retrieval, comparison, analysis, and decision-making.

6.4.1 Applications of IAM and Associated Work Processes

This section outlines a few examples of applications of IAMod workflows and practices and their transformation to management and optimization of large-scale assets. The primary objective of an IAM process is to formulate an optimized, cost-effective development plan. This objective is primarily achieved through the integration of realistic and variable production profiles and taking into account the impact of system backpressure and real-time changes in the operating conditions. Moreover, the IAM practice acts to quantify and reduce uncertainty in the design data in terms of planned production for facility maintenance and future upgrading and replacements. Last but not the least, the IAM framework provides a platform for an efficient and timely production optimization under different development schemes. When an IAMod workflow is deployed within a broader framework that involves monitoring, analysis, and decision-making to optimize business-performance results and key indicators, the IAMod transforms the operation to integrated asset management (IAM).

In Al Marzouqi et al. (2016), the IAM framework is defined as a "collection of building blocks, processes, and workflows from surveillance strategy to opportunity generation and execution monitoring." The IAM is not only the simulation technologies discussed so far in this section, but also the multidiscipline interactions and processes that use the technology effectively for decision-making. Thus, based on the experience and lessons learned in implementing IAM practices across an operator's assets in a 2-year period, Al Marzouqi et al. (2016) present a complex framework of 6 blocks, 18 processes, and 29 workflows. The operational benefits are seen through shortening the learning curves of new employees, increased productivity, more efficient key performance indicator (KPI) validation, and improved consistency of deliverables, where process standardization plays a major role. In this analogy, for example, the implementation of detailed responsible-accountable-consulted-informed (RACI) charts/matrices is warranted to properly manage human potential and understand complex relationships and governance across the organization and disciplines.

This type of effort is required to assign stakeholders who are responsible for the execution of assignments, to assure accountability, and to secure sustainability of collaborative DOF projects.

Many operators as well as service providers consider data IAM architecture as another pillar in support of DOF assets. For end users, the IAM framework should provide a transparent, easy-to-use user interface for defining and executing a variety of workflows from reservoir simulation to economic evaluation and optimization (Soma et al., 2006). At the same time, from the software implementation perspective, the IAM framework should facilitate the seamless interaction of diverse and independent applications that are responsible for various tasks in the workflow. To emphasize the relevance of data architecture, Soma et al. (2006) and Kozman (2004) highlight data composition, abstraction and federation, and visual aggregation as vital building blocks in a service-oriented IAM architecture.

Khedr et al. (2009) present the IAM workflow for optimizing large-scale artificial lift (AL) and enhanced oil recovery (EOR) strategies to one of the largest offshore fields in the world, covering an areal extent of $1200\,km^2$. The field combines three major reservoirs that produce from approximately 450 single- and dual-string wells. The field was initially developed with peripheral waterflooding strategy, and then converted into a five-spot pattern water–injection scheme. The development plan combines intensive infill drilling and applications of AL and EOR to different reservoir areas, including water injection (WI), water alternate gas (WAG), and gas injection (GI).

The deployed IAM platform couples industry-standard, domain-specific reservoir and surface network simulation applications using an explicit network balancing algorithm (Ghorayeb et al., 2003) that is well suited for solving optimization problems with a large number of wells (>500). The IAM coupling platform uses a general-purpose workflow process controller (WPC) that supports various coupling schemes, where the purpose of a process controller is to keep an external network balanced with reservoir simulation(s) as the reservoir conditions evolve:

- *Tight, iteratively lagged coupling scheme*, when the network couples to a single reservoir model: The coupling points may be individual well tubing heads or well groups. The simulator determines the pressure drop from the well bottom hole to the tubing head using the precalculated VFP tables. The list of coupling points may be extended to include the well bottom hole. Fig. 6.12 shows an example of generic tight, iterative coupling workflow. When interfacing with parallel and distributed computing systems, the controller carries dual functionality:

- Constructs the message packets and forwards them to the distributed system annex which contains a set of parallel virtual machine (PVM) system calls to communicate with the running PVM daemon.
- Communicates with the network and reservoir simulator through an open interface, which contains the PVM connectivity and controls the communication between the host computers.

Fig. 6.13 shows an example of a tight coupling architecture extended with PVM connectivity.

- *Loose-coupling scheme*, at specific time intervals, when two or more reservoir simulators are coupled to a network with common global constraints: The workflow allocates global production and injection targets to the principal groups of surface networks at the start of each synchronization step. When the synchronization steps overlap with the reservoir simulation time-steps, the loose-coupling scheme becomes an explicit-coupling scheme.
- *Explicit coupling*, when the balancing between reservoir simulator and surface network is performed exactly at the start of each simulation time-step. Fig. 6.10 above shows an example of an explicit-coupling scheme.

The IAM workflow was deployed and extensively validated in five scenarios (Khedr et al., 2009) according to the business plan constraints that mandate

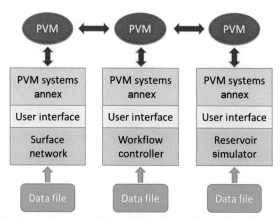

Fig. 6.13 Architecture of the coupled reservoir simulation and surface network system with PVM connectivity. *(Modified from Ghorayeb, K., Holmes, J., Torrens, R., Grewal, B., 2003. A General Purpose Controller for Coupling Multiple Reservoir Simulations and Surface Facility Networks. SPE-79702-MS. https://doi.org/10.2118/79702-MS.)*

an increase of 50% in daily field production over a 6-year period. When the increase has been achieved, the requirement is to maintain the plateau for 25 years before the field is allowed to go into decline. The validation process demonstrated that the IAM approach to modeling subsurface and surface components is crucial, particularly when more than one reservoir with different potentials share a common surface facility. For example, the study indicates that the large-scale WAG/GI application to wells producing through wellhead towers and extensive pipeline networks would be impractical to maintain the plateau. On the other hand, a new island-concept facility design represents a viable and sustainable mitigation strategy. The results of the IAM process have thus helped to make better decisions and reduce risk for commissioning a new facility layout and optimize the AL and displacement mechanisms.

Recently, Carvajal et al. (2010) introduced a holistic automated workflow for reservoir production optimization. They deploy a stochastic MIGA optimization algorithm and propose a novel definition of the optimum point, a so-called holistic point, which represents an enhancement of a traditional global optimum point: while the latter corresponds to optimization design where one or more modes can violate optimization constraints, the holistic optimum point honors all optimization constraints and renders no violations. The holistic optimization workflow is applied to two optimization problems: (a) fracture modeling and optimization (FMO), where multiobjective function is defined to maximize NPV and gas production rate and to minimize the fracture job costs and (b) reservoir development under uncertainty (RDU), where the optimization corresponds to a history matching problem to minimize the misfit in field-level reservoir pressure by parameterizing aquifer properties, such as aquifer permeability and thickness, outer/inner radius, and water encroachment angle. The schematic rendering of FMO and RDU holistic optimization workflows is given in Fig. 6.14.

The holistic FMO workflow integrates fracture design with well and reservoir modeling and clearly demonstrates the holistic optimal point in approx. 120 iterations of MIGA optimization. The holistic RDU workflow integrates material balance, well, and reservoir simulation in uncertainty quantification scheme, does not reach a holistic optimal point, despite performed 2600 MIGA iterations; however, it still captures comprehensive risk and uncertainty information and increases the confidence in production profiles for the field development plan and initial reserves estimation.

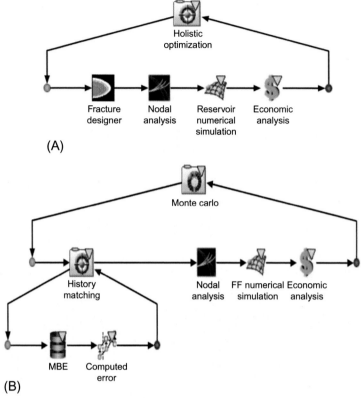

(A)

(B)

Fig. 6.14 Schematic rendering of holistic optimization workflow as designed for (A) FMO and (B) RDU. *(With permission from Carvajal, G.A., Toro, M., Szatny, M., Robinson, G., Estrada, J., 2010. Holistic Automated Workflows for Reservoir and Production Optimization. SPE-130205-MS. https://doi.org/10.2118/130205-MS.)*

Strobel et al. (2012) further advance the holistic optimization approach to enable agile and responsive automation of hydraulic fracture design study using the Microsoft Upstream Reference Architecture (MURA) initiative to standardize IT tasks while Carvajal et al. (2016) develop an advanced holistic workflow for the optimization of complex fracture network (CFN) properties by integrating production data profiles, standard well completion and stimulation properties, micro-seismic information and real-time information from fiber optics such as distributed temperature sensors (DTS) and distributed acoustic sensors (DAS) to estimate the geomechanical parameters that affect the fracture geometry and productivity.

6.4.2 Challenges and Ways Forward

In the last decade, IAM frameworks have evolved from being a promising approach for the systematic management of oil and gas assets to sophisticated and complex workflows that have delivered value in assets worldwide, facilitating high-level optimization and decision support in real-time operational domains. However, IAM process and workflows may involve considerable engineer time, human resources, and computer services, and thus it may take several weeks to find a global optimum solution. Bottom-line, there are several types of optimization techniques but these may not be completely accessible or unified and thus require a lot of knowledge before implementing. With active expansion and implementation of DOF operations, IAM practices are now transforming into computationally and operationally intensive environments that involve "continuous series of decisions based on multiple criteria including safety, environmental policy, component reliability, efficient capital and operating expenditures and revenue" (Zhang et al., 2006).

However, challenges still exist that must be addressed before IAM workflows become mainstream components of DOF operations. Here's what we see as the top five challenges:

1. The engineering and management organizations need to efficiently adapt the advanced modeling and analysis IAMod environments where the relevant groups and teams will have access to all the data and models at all times.

2. The IAM framework requires a substantial effort and multidisciplinary expertise to develop high quality and accurate models for the field life cycle. Organizations must validate the value of IAM investment and resource utilization toward field performance. Most of the oil and gas operating companies are still overly compartmentalized in their operational and business models to facilitate an efficient execution of IAM workflows. As emphasized in Cosentino (2001), the process of integrating different disciplines to perform an integrated reservoir study, requires a "change of focus." An example of a successful IAM practice that leverages and deploys the studies decision synergy (SDS) can be found in Elrafie et al. (2010). This practice represents an efficient unification platform that improves the quality of reservoir modeling through the cross-discipline integration of geology, characterization, engineering, history matching, and prediction. The benefits from its execution are gained through improved reserve quantification and optimized field development planning decisions.

3. The data required to build integrated asset models are often stored in disparate locations, are in structured and unstructured formats, are not

linked to a common database, span long production and acquisition times, and have not been adequately and consistently quality checked.

4. For large-scale IAM projects, the reservoir simulation run times can still represent a major bottleneck. The fine-scale integrated reservoir simulation models can easily exceed 100 million grid cells. Even when upscaled, the model size usually remains well above a few tens of millions grid cells. As recently indicated in Maucec et al. (2017), an ensemble-based uncertainty quantification (UQ) and AHM workflow alone, embedded in the large-scale IAM project (grid-size: approximately 34 million cells; 24 uncertainty scenarios; production period: approximately 55 years) can easily require the simulation model to run sequentially over 20–30 h, using thousands of CPU cores. Ideally, the UQ-IAM workflows in large-scale IAM projects will be executed on dedicated HPC frameworks, which in practice can be frequently difficult to achieve.

5. Last but not the least, demonstrating the value of IAM application in large-scale mature fields can sometimes be quite challenging. IAM in mature fields can be considered to have less analytical value because of lowering operating surface pressures, already existing facilities, known well performance, and well-described subsurface geology. However, when applied to mature fields, the IAM workflows can aid in more accurate estimation of the remaining reserves, reduce risks through better understanding of the interaction between subsurface and surface, indicate real opportunities for optimization scenarios, and facilitate better decisions for field development planning or well operations.

The push toward the DOF of the future will require a new generation of IAM workflows that seamlessly integrate the plethora of software tools for modeling, simulation, prediction, and optimization of an asset's performance. Moreover, the next-generation IAM must be able to operate under the extreme conditions of integrating unprecedented complexity of data in terms of the volume and transfer speed. And finally, teams will have to organize in new ways around the processes.

We have investigated several IAM concepts that may be considered the future of IAM workflows for DOF applications. Zhang et al. (2006) describe the model-based framework for oil production forecasting and optimization. Some of the key objectives of the proposed design are: generic, reusable, and flexible IAM framework architecture; adoption of a variety of assets and workflows; centralization and transparency of a common database; unification of information that handles the disparity in data formats; tool integration through loose coupling; and standards-based implementation. Zhang et al. (2006) propose the use of domain-specific modeling language based on Unified Modeling Language (UML), which provides a common, standardized vocabulary for domain experts to define and understand the asset model. In the framework of UML, they propose a template tool to provide the scenario management with a mechanism to automatically configure and invoke the forecasting tool for a particular scenario with underlying uncertainty modeling.

Bravo et al. (2011) further acknowledge that the challenges in resource negotiation, ineffective communication language, and delayed decision-making protocols can have a deteriorating effect on the execution of today's IAM workflows. They propose addressing these challenges by implementing distributed artificial intelligence (AI)-based architecture, designed for automated production management, which they call an integrated production management architecture (IPMA). The IPMA framework has three layers (see Fig. 6.15):

- *Connectivity layer*: defines data acquisition, treatment, and interpretation mechanisms.
- *Semantic layer*: consists of an ontological framework that facilitates effective information interchange between production applications. An ontological framework provides a robust and evolving vocabulary that

> **Management layer**
> - Business process automation
> - Intelligent workflow automation
> - Oilfield multiagent systems (MAS)

> **Semantic layer**
> - Production data ontology (states, relations)
> - Data meta model (objects, events, methods)

> **Connectivity layer**
> - Information sources access (production, economic database)
> - Web services
> - Optimization systems

Fig. 6.15 Integrated production management architecture (IPMA). *(Modified from Bravo, C., Saputelli, L., Castro. J.A., Rios, A., Rivas, F., Aguilar-Martin, J., 2011. Automation of the Oilfield Asset via an Artificial Intelligence (AI)-Based Integrated Production Management Architecture (IPMA). SPE-144334-MS. https://doi.org/10.2118/144334-MS.)*

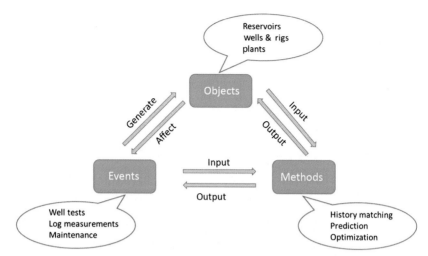

Fig. 6.16 Example of an oil production data ontology definition: generic framework of relationships.

combines terms (or nodes) and relationships that connect the terms/ nodes. An ontological framework is usually represented in the form of a directed acyclic graph. Ontology evolves with new knowledge and technology and poses no need for update schema, database revisions, etc. Fig. 6.16 shows an example of a production data ontology framework.

- *Management layer*: defines intelligent workflow automation mechanisms based on AI techniques and oilfield multiagent systems (MAS) (Bravo et al., 2011) that describe the dialog relations between processes, production unit agents, and service agents.

Bravo et al. (2011) deploy and validate the IPMA workflow on a virtual oilfield, based on a commercial integrated production model and history-matched data. The experimental oilfield has three reservoirs, eight oil wells and one flow station. The simulation results proved the efficient integral performance of IPMA layers and provided an effective solution to the production optimization problem. The benefits of IPMA architecture are seen through:

- Standardized mechanism to communicate the process state information sources.
- Introduction and formulation of an oil and gas production data ontology for the information exchange between production applications.
- Decrease in information search time.
- Deployment of a flexible mechanism for business-process automation.

REFERENCES

Al Marzouqi, M.A., Bahamaish, J., Al Jenaibi, H., Al Hammadi, H., Saputelli, L., 2016. Building ADNOC's Pillars for Process Standardization and Best Practices through an Integrated Reservoir Management Framework. SPE-183421-MS, https://doi.org/10.2118/183421-MS.

Allen, D.P.B., 2017. Handling Risk and Uncertainty in Portfolio Production Forecasting. SPE-185178-PA, https://doi.org/10.2118/185178-PA.

Alpak, F.O., Vink, J.C., Gao, G., Mo, W., 2013. Techniques for Effective Simulation, Optimization and Uncertainty Quantification of the In-Situ Upgrading Process. SPE-163665-MS, https://doi.org/10.2118/163665-MS.

Alpak, F.O., Jin, L., Ramirez, B.A., 2015. Robust Optimization of Well Placement in Geologically Complex Reservoirs. Paper SPE 175106-MS, https://doi.org/10.2118/175106-MS.

Al-Shamma, B.R., Gosselin, O., King, P., Christie, M., Mendez, M., 2015. Comparative Performance of History Matching Algorithms Using Diverse Parameterization Methods: Application to a Synthetic and North Sea Cases. SPE-174360-MS, https://doi.org/10.2118/174360-MS.

April, J., Glover, F., Kelly, J., Laguna, M., Erdogan, M., Mudford, B., Stegemeier, D., 2003. Advanced Optimization Methodology in the Oil and Gas Industry: The Theory of Scatter Search Techniques With Simple Examples. SPE-82009-MS, https://doi.org/10.2118/82009-MS.

Arnold, D., Demyanov, V., Christie, M., Gopa, K., Bakay, A., 2016. Optimisation of decision making under uncertainty throughout field lifetime. J. Comput. Sci. Geosci. 95 (C), 123–139. http://dl.acm.org/citation.cfm?id=2994918.

Arroyo, E., Devegowda, D., Datta-Gupta, A., Choe, J., 2008. Streamline-Assisted Ensemble Kalman Filter for Rapid and Continuous Reservoir Model Updating. SPE-104255-PA, https://doi.org/10.2118/104255-PA.

Bailey, W.J., Couet, B., Wilkinson, D., 2004. Field Optimization Tool for Maximizing Asset Value. SPE-87026-MS, https://doi.org/10.2118/87026-MS.

Barros, E.G.D., Leeuwenburgh, O., van den Hof, P.M.J., Jansen, J.D., 2015. Value of Multiple Production Measurements and Water Front Tracking in Closed-Loop Reservoir Management. SPE-175608-MS, https://doi.org/10.2118/175608-MS.

Basu, S., Das, A., De Paola, G., Echeverria Ciaurri, D., Droz, S.E., Llano, C.I., Ocheltree, K.B., Rodriguez, R., 2016. Multi-start Method for Reservoir Model Uncertainty Quantification With Application to Robust Decision Making. IPTC-18711-MS, https://doi.org/10.2523/IPTC-18711-MS.

Beckner, B.L., Song, X., 1995. Field development planning using simulated annealing—optimal economic well scheduling and placement. In: SPE 30650.

Begg, S.H., Bratvold, R.B., Campbell, J.M., 2001. Improving Investment Decisions Using a Stochastic Integrate Asset Model. SPE-71414-MS, https://doi.org/10.2118/77586-MS.

Bhark, E., Rey, A., Datta-Gupta, A., Jafarpour, B., 2011. Multiscale Parameterization and History Matching in Structured and Unstructured Grid Geometries. SPE-141764-MS, https://doi.org/10.2118/141764-MS.

Bieker, H.P., Sluphaugg, O., Johansen, T.A., 2006. Real Time Production Optimization of Offshore Oil and Gas Production Systems: A Technology Survey. SPE-99446-MS, https://doi.org/10.2118/99446-MS.

Bouzarkouna, Z., Ding, D.Y., Auger, A., 2013. Partially Separated Metamodels with Evolution Strategies for Well-Placement Optimization. SPE-143292-PA, https://doi.org/10.2118/143292-PA.

Bratvold, R.B., Begg, S.H., Rasheva, S., 2010. A New Approach to Uncertainty Quantification for Decision Making. SPE-130157-MS, https://doi.org/10.2118/130157-MS.

Bravo, C., Saputelli, L., Castro, J.A., Rios, A., Rivas, F., Aguilar-Martin, J., 2011. Automation of the Oilfield Asset via an Artificial Intelligence (AI)-Based Integrated Production Management Architecture (IPMA). SPE-144334-MS, https://doi.org/10.2118/144334-MS.

Brouwer, D.R., Jansen, J.D., 2002. Dynamic Optimization of Water Flooding with Smart Wells using Optimal Control Theory. SPE-78278-MS, https://doi.org/10.2118/78278-MS.

Brouwer, D.R., Nævdal, G., Jansen, J.D., Vefring, E.H., van Kruisdijk, C.P.J.W., 2004. Improved Reservoir Management through Optimal Control and Continuous Model Updating. SPE-90149-MS, https://doi.org/10.2118/90149-MS.

Carvajal, G.A., Toro, M., Szatny, M., Robinson, G., Estrada, J., 2010. Holistic Automated Workflows for Reservoir and Production Optimization. SPE-130205-MS, https://doi.org/10.2118/130205-MS.

Carvajal, G.A., Boisvert, I., Knabe, S., 2014. A Smartflow for SmartWells: Reactive and Proactive Modes. SPE-167821-MS, https://doi.org/10.2118/167821-MS.

Carvajal, G.A., Khoriakov V., Filppov, A., Maucec, M., Diaz, A., Knabe, S., 2016. Estimating Well Production Performance in Fractured Reservoir Systems. US Patent application #20160259088.

Chen, C., Li, G., Reynolds, A.C., 2012. Robust Constrained Optimization of Short- and Long-term Net Present Value for Closed-Loop Reservoir Management. SPE-141314-PA, https://doi.org/10.2118/141314-PA.

Chen, C., Gao, G., Honorio, J., Gelderblom, P., Jimenez, E., Jaakkola, T., 2014. Integration of Principal Component Analysis and Streamline Information for the History Matching of Channelized Reservoirs. SPE-170636-MS, https://doi.org/10.2118/170636-MS.

Chen, Y., Oliver, D.S., 2013. Levenberg-Marquardt forms of the iterative ensemble smoother for efficient history matching and uncertainty quantification. Comput. Geosci. 17 (4), 689–703.

Chen, Y., Oliver, D.S., Zhang, D., 2009. Efficient Ensemble-based Closed-Loop Production Optimization. SPE 112873-PA, https://doi.org/10.2118/112873-MS.

Chow, C.V., Arnondin, M.C., 2000. Managing risks using integrated production models: process description. J. Pet. Technol. (March), 54, https://doi.org/10.2118/57472-JPT.

Compan, A.L.M., Bodstein, G.C.R., Couto, P., 2016. A Relative Permeability Rock-Typing Methodology with a Clustering Method Combined with a Heuristic Optimization Procedure. SPE-180916-PA, https://doi.org/10.2118/180916-PA.

Cope, G., 2011. Optimizing Optimization: New software technology is revolutionizing reservoir simulation. New Technology Magazine, November 2011. pp. 15–19.

Cosentino, L., 2001. Integrated Reservoir Studies. IFP Publications, Editions,TECHNIP.

Cullick, A.S., Johnson, D., Shi, G., 2006. Improved and More Rapid History Matching with a Nonlinear Proxy and Global Optimization. SPE-101933-MS, https://doi.org/10.2118/101933-MS.

Cullick, A.S., Heath, D., Narayanan, K., April, J., Kelly, J., 2003. Optimizing Multiple-Field Scheduling and Production Strategy with Reduced Risk. SPE-84239-MS, https://doi.org/10.2118/84239-MS.

Cullick, A.S., Heath, D., Narayanan, K., April, J., Kelly, J., 2004. Optimizing Multiple-Field Scheduling and Production Strategy with Reduced Risk. SPE-88991-JPT, https://doi.org/10.2118/88991-JPT.

Deutsch, C.V., Journel, A.G., 1994. The Application of Simulated Annealing to Stochastic Reservoir Modeling. SPE Adv. Technol. Ser. 2 (2), 222–227.

Dilib, F.A., Jackson, M.D., 2013. Closed-Loop Feedback Control for Production Optimization of Intelligent Wells under Uncertainty. SPE-112873-PA, https://doi.org/10.2118/150096-PA.

Dzuyba, V.I., Bogachev, K.Y., Bogaty, A.S., Lyapin, A.R., Mirgasimov, A.R., Semenko, A.E., 2012. Advances in Modeling of Giant Reservoirs. SPE-163090-MS, https://doi.org/10.2118/163090-MS.

Echeverria Ciaurri, D., Mukerji, T., Durlofsky, L.J., 2011. Derivative-free Optimization for Oil Field Operations. In: Computational Optimization and Applications in Engineering and Industry. Studies in Computational Intelligence, vol. 359. Springer, Berlin, Heidelberg, Germany, pp. 19–55.

Echeverria Ciaurri, D., Conn, A.R., Mello, U.T., Onwunalu, J.E., 2012. Integrating Mathematical Optimization and Decision Making in Intelligent Fields. SPE-149780-MS, https://doi.org/10.2118/149780-MS.

Eeg, O.S., Herring, T., 1997. Combining Linear Programming and reservoir Simulation to Optimize Asset Value. SPE-37446-MS, https://doi.org/10.2118/37446-MS.

Efendiev, Y., Datta-Gupta, A., Ginting, V., Ma, X., Mallick, B., 2005. An Efficient Two-Stage Markov Chain Monte Carlo Method for Dynamic Data Integration. https://doi.org/10.1029/2004WR003764.

Elrafie, E.A., Hogg, M., Mohammadi, H.H., 2010. Integrated Reservoir Studies Roll-Up Initiative—A New Industry Step Change Innovation. SPE-138551-MS, https://doi.org/10.2118/138551-MS.

Emerick, A.A., Reynolds, A.C., 2012. History-matching Time Lapse Seismic Data Using Ensemble Kalman Filter with Multiple Data Assimilation. Comput. Geosci. 16 (3), 639–659. https://link.springer.com/article/10.1007/s10596-012-9275-5.

Emerick, A.A., Reynolds, A.C., 2013. Ensemble Smoother with Multiple Data Assimilation. Comput. Geosci. 55, 3–15.

Evensen, G., 1994. Sequential data assimilation with a nonlinear quasi-geostrophic model using Monte Carlo methods to forecast error statistics. J. Geophys. Res. 99 (C5), 10143–10162.

Evensen, G., 2009. Data Assimilation: The Ensemble Kalman Filter, second ed. Springer Verlag, Berlin, Germany.

Ferraro, P., Verga, F., 2009. In: Use of Evolution Algorithms in Single- and Multi-Objective Optimization Techniques for Assisted History matching.Paper Presented at the Offshore Mediterranean Conference and Exhibition, Ravenna, Italy, 25–27 March, 2009.

Fillacier, S., Fincham, A.E., Hammersley, R.P., Heritage, J.R., Kolbikova, I., Peacock, G., Soloviev, V.Y., 2014. Calculating Prediction Uncertainty using Posterior Ensemble Generated from Proxy Models. SPE-171237-MS, https://doi.org/10.2118/171237-MS.

Fleming, G., Wang, Q., 2017. A Parallel Solution for Large Surface Networks in a Fully Integrated Reservoir Simulator. Paper SPE-182634-MS, https://doi.org/10.2118/182634-MS.

Floudas, C.A., Pardalos, P.M. (Eds.), 2009. Encyclopedia of Optimization. Springer, USA. ISBN 978-0-387-74760-6.

Fu, J., Wen, X.-H., 2017. Model-Based Multi-Objective Optimization Methods for Efficient Management of Subsurface Flow. SPE-182598-MS, https://doi.org/10.2118/182598-MS.

Ghorayeb, K., Holmes, J., Torrens, R., Grewal, B., 2003. A General Purpose Controller for Coupling Multiple Reservoir Simulations and Surface Facility Networks. SPE-79702-MS, https://doi.org/10.2118/79702-MS.

Gomez, Y., Khazaeni, Y., Mohaghegh, S.D., Gaskari, R., 2009. Top-Down Intelligent Reservoir Modeling. SPE-124204-MS, https://doi.org/10.2118/124204-MS.

Goodwin, N., 2015. Bridging the Gap Between Deterministic and Probabilistic Uncertainty Quantification Using Advanced Proxy Based Methods. SPE-173301-MS, https://doi.org/10.2118/173301-MS.

Goodwin, N., Esler, K., Ghasemi, M., Mukundakrishnan, K., Wang, H., Gilman, J.R., Lee, B., 2017. Probabilistic Uncertainty Quantification of a Complex Field using Advanced Proxy Based Methods and GPU-Based Reservoir Simulation. SPE-182637-MS, https://doi.org/10.2118/182637-MS.

Hajizadeh, Y., 2011. Population-Based Algorithms for Improved History Matching and Uncertainty Quantification of Petroleum Reservoirs. PhD Thesis, Heriot Watt University, Edinburgh, Scotland, p. 331.

Hajizadeh, Y., Christie, M., Demyanov, V., 2009. Ant Colony Optimization for History Matching. SPE-121193-MS, https://doi.org/10.2118/121193-MS.

Hajizadeh, Y., Christie, M., Demyanov, V., 2010. History Matching with Differential Evolution Approach; a Look at New Search Strategies. SPE-130253-MS, https://doi.org/10.2118/130253-MS.

Hanea, R., Evensen, G., Husoft, L., Ek, T., Chitu, A., Wilschut, F., 2015. Reservoir Management under Geological Uncertainty using Fast Model Update. SPE-173305-MS, https://doi.org/10.2118/173305-MS.

Hegstad, B.K., Sætrom, J., 2014. Using Multiple Realizations from an Integrated Uncertainty Analysis to Make More Robust Decisions in Field Development. SPE-171831-MS, https://doi.org/10.2118/171831-MS.

Honorio, J., Chen, C., Gao, G., Du, K., Jaakkola, T., 2015. Integration of PCA with Novel Machine Learning Method for Parameterization and Assisted History Matching Geologically Complex Reservoirs. SPE-175038-MS, https://doi.org/10.2118/175038-MS.

Hutahaean, J., Demyanov, V., Christie, M.A., 2017. On Optimal Selection of Objective Grouping for Multi-Objective History Matching. SPE-185957-PA, https://doi.org/10.2118/185957-PA.

Hutahaean, J.J., Demyanov, V., Christie, M., 2015. Impact of Model Parameterization and Objective Choices on Assisted History Matching of Reservoir Forecasting. SPE-176389-MS, https://doi.org/10.2118/176389-MS.

Isebor, O.J., Echeverria Ciaurri, D., Durlofsky, L.J., 2014. Generalized Field-Development Optimization with Derivative-Free Procedures. SPE-163631-PA, https://doi.org/10.2118/163631-PA.

Iyer, R.R., Grossmann, I.E., Vasantharajan, S., Cullick, A.S., 1998. Optimal planning and scheduling offshore oil field infrastructure, investment and operations. Ind. Eng. Chem. Res. 37 (4), 1380–1397. https://doi.org/10.1021/ie970532x.

Jafarpour, B., McLaughlin, D.B., 2007. Efficient Permeability Parameterization with the Discrete Cosine Transform. SPE-106453-MS, https://doi.org/10.2118/106453-MS.

Jafarpour, B., McLaughlin, D.B., 2009. Estimating Channelized-Reservoir Permeabilities with the Ensemble Kalman Filter: The Importance of Ensemble Design. SPE-108941-PA, https://doi.org/10.2118/108941-PA.

Jansen, J.D., Douma, S.D., Brouwer, D.R., van den Hof, P.M.J., Bosgra, O.H., Heemink, A.W., 2009. Closed-Loop Reservoir Management. SPE-119098-MS, https://doi.org/10.2118/119098-MS.

Jesmani, M., Bellout, M.C., Hanea, R., Foss, B., 2015. Particle Swarm Optimization Algorithm for Optimum Well Placement Subject to Realistic Field Development Constraints. SPE-175590-MS, https://doi.org/10.2118/175590-MS.

Jin, L., Villa, J., Webber, D., van den Hoek, P., Pirmez, C., 2012. A Novel Workflow in Experimental Design Framework Integrating Selected 4D Seismic Measurements into Reservoir Simulation Models. SPE-159136-MS, https://doi.org/10.2118/159136-MS.

John, U.M., Onjekonwu, M.O., 2010. Non-Linear Programming of Well Spacing Optimization of Oil Reservoirs. SPE-140674-MS, https://doi.org/10.2118/140674-MS.

Kam, D., Han, J., Datta-Gupta, A., 2016. Streamline-based Rapid History Matching of Bottomhole Pressure and Three-phase Production Data. SPE-179549-MS, https://doi.org/10.2118/179549-MS.

Kang, B., Lee, K., Choe, J., 2015. Efficient Sampling Scheme for Uncertainty Quantification Using PCA. SPE-176183-MS, https://doi.org/10.2118/176183-MS.

Kang, S., Bhark, E., Datta-Gupta, A., Kim, J., Jang, I., 2014. A Hierarchical Model Calibration Approach with Multiscale Spectral-Domain Parameterization: Application to a Structurally Complex Fractured Reservoir. SPE-169061-MS, https://doi.org/10.2118/169061-MS.

Kansao, R., Yrigoyen, A., Haris, Z., Saputelli, L., 2017. Waterflood performance diagnosis and optimization using data-driven predictive analytical techniques from capacitance resistance models CRM. In: SPE 185813-MS. https://doi.org/10.2118/185813-MS.

Khedr, O., Al Marzouqi, M., Torrens, R., Amtereg, A., 2009. On the Importance and Application of Integrated Asset Modeling of a Giant Offshore Oil Field. SPE-123689-MS, https://doi.org/10.2118/123689-MS.

Khedr, O.H., Amur, J., Al-Ameri, R., Torrens, R., Amtereg, A.A., 2012. A Unique Integrated Asset Modeling Solution to Optimize and Manage Uncertainty in a Giant Offshore Oil Field Development Mega-Project. SPE-161280-MS, https://doi.org/10.2118/161280-MS.

Kitanidis, P.K., 1995. Quasi-linear geo-statistical theory for inversion. Water Resour. Res. 31 (10), 2411–2419.

Klie, H., 2015. Physics-Based and Data-Driven Surrogates for Production Forecasting. SPE-173206-MS, https://doi.org/10.2118/173206-MS.

Kozman, J.B., 2004. Why Can't I just Start with a Map?—Case Histories for Integrated Asset Management. SPE-87022-MS, https://doi.org/10.2118/87022-MS.

Kumar, S., Wen, X.-H., He, J., Lin, W., Yardumian, H., Fahruri, I., et al., 2017. Integrated Static and Dynamic Uncertainty Modeling Big-Loop Workflow Enhances Performance Prediction and Optimization. SPE-182711-MS, https://doi.org/10.2118/182711-MS.

Le, D.H., Emerick, A.A., Reynolds, A.C., 2015. An adaptive ensemble smoother with multiple data assimilation for assisted history matching. In: SPE-173214-MS. https://doi.org/10.2118/173214-MS.

Li, J., Jiang, H., Liang, B., Zhou, D., Ding, S., Gong, C., Zhao, L., 2016. Injection Allocation in Multi-Layer Water Flooding Reservoirs Using SVM Optimized by Genetic Algorithm. IPTC-19006-MS, https://doi.org/10.2523/IPTC-19006-MS.

Li, X., Reynolds, A.C., 2017. Generation of a Proposal Distribution for Efficient MCMC Characterization of Uncertainty in Reservoir Description and Forecasting. SPE-182684-MS, https://doi.org/10.2118/182684-MS.

Liao, T.T., Stein, M.H., 2002. Evaluating Operation Strategies via Integrated reservoir Modeling. SPE-75525-MS, https://doi.org/10.2118/75525-MS.

Liu, X., Reynolds, A.C., 2015. Multiobjective Optimization for Maximizing Expectation and Minimizing Uncertainty or Risk with Application to Optimal Well Control. SPE-173216-MS, https://doi.org/10.2118/173216-MS.

Lorentzen, R.J., Berg, A., Naevdal, G., Vefring, E.H., 2006. A New Approach for Dynamic Optimization of Water Flooding Problems. SPE-99690-MS, https://doi.org/10.2118/99690-MS.

Ma, X., Al-Harbi, M., Datta-Gupta, A., Efendiev, Y., 2006. A Multistage Sampling Method for Rapid Quantification of Uncertainty in History Matching Geological Models. SPE-102476-MS, https://doi.org/10.2118/102476-MS.

Mata-Lima, H., 2011. Evaluation of the objective functions to improve production history matching performance based on fluid flow behavior in reservoirs. J. Pet. Sci. Eng. 78 (1), 42–53. https://doi.org/10.1016/j.petrol.2011.05.015.

Maucec, M., 2016. Systems and Methods for Generating Updates of Geological Models. US Patent 9,330,064 B2.

Maucec, M., Cullick, A.S., 2015. Systems and Methods for the Quantitative Estimate of Production-Forecast Uncertainty. US Patent 9,223,042 B2.

Maucec, M., Douma, S., Hohl, D., Leguijt, J., Jimenez, E.A., Datta-Gupta, A., 2007. Streamline-Based History Matching and Uncertainty: Markov-chain Monte Carlo Study of an Offshore Turbidite Oil Field. SPE-109943-MS, https://doi.org/10.2118/109943-MS.

Maucec, M., Cullick, S., Shi, G., 2011. Geology-guided Quantification of Production-Forecast Uncertainty in Dynamic Model Inversion. SPE-146748-MS, https://doi.org/10.2118/146748-MS.

Maucec, M., Singh, A., Carvajal, G., Mirzadeh, S., Knabe, S., Chambers, R., et al., 2013a. Engineering Workflow for Probabilistic Assisted History Matching and Production Forecasting: Application to a Middle East Carbonate Reservoir. SPE-165980-MS, https://doi.org/10.2118/165980-MS.

Maucec, M., Singh, A., Carvajal, G., Mirzadeh, S., Knabe, S., Mahajan, A., et al., 2013b. Next Generation of Workflows for Multi-level Assisted History Matching and Production Forecasting: Concept, Collaboration and Visualization. SPE-167340-MS, https://doi.org/10.2118/167340-MS.

Maucec, M., De Matos Ravanelli, F.M., Lyngra, S., Zhang, S.J., Alramadhan, A.A., Abdelhamid, O.A., Al-Garni, S.A., 2016. Ensemble-based Assisted History Matching with Rigorous Uncertainty Quantification Applied to Naturally Fractured Carbonate Reservoir. SPE-181325-MS, https://doi.org/10.2118/181325-MS.

Maucec, M., Awan, A., Benedek, L., Lyngra, S., 2017. Implementation of Assisted History Matching Under Uncertainty in Integrated Reservoir Modeling. SPE-188049-MS, https://doi.org/10.2118/188049-MS.

McVay, D.A., Dossary, M.N., 2014. The Value of Assessing Uncertainty. SPE-160189-PA, https://doi.org/10.2118/160189-PA.

Mohaghegh, S.D., Abdulla, F., Abdou, M., Gaskari, R., Maysami, M., 2015. Smart Proxy: An Innovative Reservoir Management Tool; Case Study of a Giant Mature Oilfield in UAE. SPE-177829-MS, https://doi.org/10.2118/177829-MS.

Mohamed, L., Christie, M., Demyanov, V., 2010a. Comparison of Stochastic Sampling Algorithms for Uncertainty Quantification. SPE-119139-PA, https://doi.org/10.2118/119139-PA.

Mohamed, L., Christie, M., Demyanov, V., 2010b. Reservoir Model History Matching with Particle Swarms: Variants Study. SPE-129152-MS, https://doi.org/10.2118/129152-MS.

Nævdal, G., Johnsen, L.M., Aanonsen, S.I., Vefring, E., 2003. Reservoir monitoring and continuous model updating using the ensemble Kalman filter. In: SPE-84372-MS. https://doi.org/10.2118/84372-PA.

Neal, R.M., 1993. Probabilistic Inference using Markov Chain Monte Carlo Methods. Technical Report CRG-TR-93-1, Dept. of Computer Science, University of Toronto.

Nemirovski, A., Juditsky, A., Lan, G., Shapiro, A., 2009. Robust Stochastic Approximation Approach to Stochastic Programming. SIAM J. Optim. 19 (4), 1574–1609 (Society for Applied and Industrial Mathematics), .

Newman, A.J., 1996. Model Reduction via Karhunen-Loeve Expansion Part I: An exposition. Institute for Systems Research and Electrical Engineering Department, University of Maryland. http://citeseerx.ist.psu.edu/viewdoc/summary?doi=10.1.1.50.6670&rank=1.

Oberwinkler, C., Stundner, M., 2005. From Real-Time Data to Production Optimization. SPE 87008-PA, https://doi.org/10.2118/87008-PA.

Olalotiti-Lawal, F., Datta-Gupta, A., 2015. A Multi-Objective Markov Chain Monte Carlo Approach for History Matching and Uncertainty Quantification. SPE-175144-MS, https://doi.org/10.2118/175144-MS.

Oliver, D.S., Chen, Y., 2011. Recent progress on reservoir history matching: a review. Comput. Geosci. 15 (1), 185–221. http://www.academia.edu/22805522/Recent_progress_on_reservoir_history_matching_a_review.

Oliver, D.S., 1996. Multiple Realizations of the Permeability Field from Well-test Data. SPE-27970-PA, https://doi.org/10.2118/27970-PA.

Oliver, D.S., He, N., Reynolds, A.C., 1996. Conditioning Permeability Fields to Pressure Data.Presented at the 5th European Conference on the Mathematics of Oil Recovery (ECMOR), Leoben, Austria, September 3–6, 1996.

Ombach, J., 2014. A Short Introduction to Stochastic Optimization. Schedae Informaticae 23, 9–20.

Onwunalu, J.E., Durlofsky, L.J., 2011. A New Well-Pattern Optimization procedure for Large-Scale Field Development. SPE-124364-PA, https://doi.org/10.2118/124364-PA.

Ouenes, A., Bhagavan, S., Bunge, P.H., Travis, B.J., 1994. Application of Simulated Annealing and Other Global Optimization Methods to Reservoir Description: Myths and Realities. SPE-28415-MS, https://doi.org/10.2118/28415-MS.

Ramirez, B.A., Joosten, G.J.P., Kaleta, M.P., Gelderblom, P.P., 2017. Model-Based Well Location Optimization—A Robust Approach. SPE-182632-MS, https://doi.org/10.2118/182632-MS.

Rwenchungura, R., Dadashpour, M., Kleppe, J., 2011. Advanced History Matching Techniques Reviewed. SPE-142497-MS, https://doi.org/10.2118/142497-MS.

Sætrom, J., Selseng, H., MacDonald, A., Kjølseth, T., Kolbjørnsen, O., 2016. Consistent Integration of Drill-Stem Test Data into Reservoir Models on a Giant Field Offshore Norway. SPE 181352-MS, https://doi.org/10.2118/181352-MS.

Salam, D.D., Gunardi, I., Yasutra, A., 2015. Production Optimization Strategy Using Hybrid Genetic Algorithm. SPE-177442-MS, https://doi.org/10.2118/177442-MS.

Salsburg, D., 2001. The Lady Tasting Tea—How Statistics Revolutionized Science in the Twentieth Century. Henry Holt & Co., New York, NY.

Sambo, C.H., Hematpour, H., Danaei, S., Herman, M., Ghosh, D.P., Abass, A., Elraies, K.A., 2016. An Integrated Reservoir Modeling and Evolutionary Algorithm for Optimizing Field Development in a Mature Fractured Reservoir. SPE 183178-MS, https://doi.org/10.2118/183178-MS.

Sampaio Pinto, M.A., Ghasemi, M., Sorek, N., Gildin, E., Schiozer, D.J., 2015. Hybrid Optimization for Closed-Loop Reservoir Management. SPE-173278-MS, https://doi.org/10.2118/173278-MS.

Saputelli, L., Nikolaou, M., Economides, M.J., 2003. Self-Learning Reservoir Management. SPE-84064-MS, https://doi.org/10.2118/84064-MS.

Saputelli, L., Nikolaou, M., Economides, M.J., 2006. Real-time reservoir management: a multi-scale adaptive optimization and control approach. Comput. Geosci. 10 (1), 61–96.

Sarma, P., Chen, W.H., 2011. Robust and Efficient Handling of Model Constraints with the Kernel-Based Ensemble Kalman. Filter. SPE-141948-MS, https://doi.org/10.2118/141948-MS.

Sarma, P., Chen, W.H., 2013. Preventing Ensemble Collapse and Honoring Multipoint Geostatistics with the Subspace EnKF/EnS and Kernel PCA Parameterization. SPE-163604-MS, https://doi.org/10.2118/163604-MS.

Sarma, P., Durlofsky, L.J., Aziz, K., 2005. Efficient Closed-Loop Production Optimization under Uncertainty. SPE-94241-MS, https://doi.org/10.2118/94241-MS.

Sarma, P., Durlofsky, L.J., Aziz, K., Chen, W.H., 2006. Efficient real-time reservoir management using adjoint-based optimal control and model updating. Comput. Geosci. 10 (1), 3–36.https://link.springer.com/article/10.1007/s10596-005-9009-z.

Sarma, P., Chen, W.H., Durlofsky, L.J., Aziz, K., 2008. Production Optimization with Adjoint Models under Nonlinear Control-State Path Inequality Constraints. SPE-99959-PA, https://doi.org/10.2118/99959-PA.

Sazonov, E., Nugaeva, A., Dobycha, B., Muryzhnikov, A., Eydinov, D., 2015. Risks and Uncertainties Evaluation of Reservoir Models as a Way to Optimal Decisions. SPE-176623-MS, https://doi.org/10.2118/176623-MS.

Scheidt, C., Caers, J., 2009. Representing Spatial Uncertainty Using Distances and Kernels. Math. Geosci. 41 (4), 397–419.

Schulze-Riegert, R., Nwakile, M., Skripkin, S., Willen, Y., 2016. Scalability and Performance Efficiency of History Matching Workflows using MCMC and Adjoint Techniques Applied to the North Sea Reservoir Case Study. SPE-180106-MS, https://doi.org/10.2118/180106-MS.

Schulze-Riegert, R.W., Ghedan, S., 2007. In: Modern Techniques for History Matching. Presented at the 9th International Forum on Reservoir Simulation, Abu Dhabi, UAE, 9–13 December 2007.

Schulze-Riegert, R.W., Axmann, J.K., Haase, O., Rian, D.T., You, Y.L., 2001. Optimization Methods for History Matching of Complex Reservoirs. SPE-66393-MS, https://doi.org/10.2118/66393-MS.

Schulze-Riegert, R.W., Krosche, M., Fahimuddin, A., Ghedan, S.G., 2007. Multi-Objective Optimization with Application to Model Validation and Uncertainty Quantification. SPE-105313-MS, https://doi.org/10.2118/105313-MS.

Self, R., Atashnezhad, A., Hareland, G., 2016. Reducing Drilling Cost by Finding Optimal Operational Parameters Using Particle Swarm Algorithm. SPE-180280-MS, https://doi.org/10.2118/180280-MS.

Shirangi, M., Durlofsky, L.J., 2015. Closed-Loop Field Development Optimization under Uncertainty. SPE-173219-MS, https://doi.org/10.2118/173219-MS.

Skjervheim, J.-A., Evensen, G., Hove, J., Vabø, J.G., 2011. An Ensemble Smoother for Assisted History Matching. SPE-141929-MS, https://doi.org/10.2118/141929-MS.

Soma, R., Bakshi, A., Orangi, A., Prasanna, V.K., Da Sie, W., 2006. A Service-Oriented Data-Composition Architecture for Integrated Asset Management. SPE-99983-MS, https://doi.org/10.2118/99983-MS.

Strobel, M.A., Carvajal, G.A., Szatny, M., Peries, C., 2012. Enabling Agile and Responsive Workflow Automation—A Hydraulic Fracture Design Case Study. SPE-150455-MS, https://doi.org/10.2118/150455-MS.

Sui, W., Mengfei, Z., Wang, X., Zhang, S., 2014. Reconstruction of Shale Using Dual-Region Strategy and Very Fast Simulated Annealing Algorithm. URTEC-1897043-MS, https://doi.org/10.15530/URTEC-2014-1897043.

Suwartadi, E., Krogstad, S., Foss, B., 2011. Nonlinear output constraints handling for production optimization of oil reservoirs. Comput. Geosci. 16 (2), 499–517. https://link.springer.com/article/10.1007/s10596-011-9253-3.

Temizel, C., Purwar, S., Urrutia, K., Abdullayev, A., Md Adnan, F., Agarwal, A., et al., 2014. Optimization of Well Placement in real-Time Production Optimization of Intelligent Fields With Use of Local and Global Methods. SPE-170675-MS, https://doi.org/10.2118/170675-MS.

Temizel, C., Purwar, S., Urrutia, K., Abdullayev, A., 2015. Real-time Optimization of Waterflood Performance through Coupling Key Performance Indicators in Intelligent Fields. SPE-173402-MS, https://doi.org/10.2118/173402-MS.

Toby, S., 2014. Making the Best of Integrated Asset Modeling. SPE-171161-MS, https://doi.org/10.2118/171161-MS.

Ursini, F., Rossi, R., Pagliari, F., 2010. Forecasting Reservoir Management through Integrated Asset Modeling. SPE-128165-MS, https://doi.org/10.2118/128165-MS.

van Essen, G.M., Jimenez, E.A., Przybysz-Jarnut, J.P., et al., 2012. Adjoint-based History Matching of Production and Time-lapse Seismic Data. SPE-154375-MS, https://doi.org/10.2118/154375-MS.

van Leeuwen, P.J., Evensen, G., 1996. Data Assimilation and Inverse Methods in Terms of a Probabilistic Formulation. Mon. Weather Rev. 124 (12), 2898–2913.

Vanderheyden, W.B., Dozzo, J.A., Mastny, E., Hanson, B., Kaesbach, I., Smart, G.T., et al., 2016. Complex Facilities and Multireservoir Production Management Using a Tightly Integrated High Performance Simulator with a Flexible User Procedure Facility. SPE-181597-MS, https://doi.org/10.2118/181597-MS.

Vasquez, M., Suarez, A., Aponte, H., Ocanto, L., Fernandes, J., 2001. Global Optimization of Oil Production Systems, A Unified Operational View. SPE-71561-MS, https://doi.org/10.2118/71561-MS.

Wang, C., Li, G., Reynolds, A.C., 2007. Production Optimization in Closed-Loop Reservoir Management. SPE-109805-MS, https://doi.org/10.2118/109805-MS.

Wang, C., Li, G., Reynolds, A.C., 2009. Production Optimization in Closed-Loop Reservoir Management. SPE 109805-PA, https://doi.org/10.2118/109805-PA.

Wang, P., Aziz, K., Litvak, M.L., 2002. Optimization of Production from Mature Fields. WPC-32152 presented at the 17th World Petroleum Congress, Rio de Janeiro, Brazil, 1–5 September 2002.

Williams, G.J.J., Mansfield, M., MacDonald, D.G., Bush, M.D., 2004. Top-Down Reservoir Modeling. SPE-89974-MS, https://doi.org/10.2118/89974-MS.

Wiluweit, M., Bin Khairul Azmi, M., Silalahi, E., Flew, S., 2015. Application of Big Loop Uncertainty Analysis to Assist History Matching and Optimize Development of a Waterflood Field. SPE-176464-MS, https://doi.org/10.2118/176464-MS.

Yanai, H., Takeuchi, K., Takane, Y., 2011. Singular value decomposition (SVD). In: Projection Matrices, Generalized Inverse Matrices, and Singular Value Decomposition. Statistics for Social and Behavioral SciencesSpringer, New York, NY.

Yang, C., Nghiem, L., Card, C., Bremeier, M., 2007. Reservoir Modeling Uncertainty Quantification through Computer-Assisted History Matching. SPE-109825-MS, https://doi.org/10.2118/109825-MS.

Zhang, C., Orangi, A., Bakshi, A., Da Sie, W., Prasanna, U., 2006. Model-Based Framework for Oil production Forecasting and Optimization: A Case Study in Integrated Asset Management. SPE-99979-MS, https://doi.org/10.2118/99979-MS.

CHAPTER SEVEN

Smart Wells and Techniques for Reservoir Monitoring

Contents

This chapter introduces concepts associated with smart well technology and its application to maximize the oil–recovery factor and improve financial indicators. In 1997, the first successful "smart well" was installed in a well in the Norwegian sector of the North Sea. What made the well "smart"? The completion incorporated permanently installed, down–hole pressure and temperature measurements integrated with remotely controlled, high–fidelity flow-control valves. Konopczynski and Ajayi (2008) stated that this event marked the genesis of the intelligent well era. (Note: "Smart well" and "intelligent well" are used interchangeably.)

Over the past decade, the use of intelligent well technology in many geographic regions has "crossed the technology adoption chasm," as oil and gas producers are increasingly using this technology in field developments to improve reservoir management, which is the main benefit that intelligent well technology delivers. Current technical challenges focus on new and modified mechanical systems for the valve configurations and alternatives such as interval control devices and the control strategies to operate the mechanics for optimal reservoir management such as water or gas breakthrough. This chapter explains technical aspects of optimizing oil production with smart wells.

7.1 INTRODUCTION TO SMART WELLS

A smart well can be defined as a well that uses mechanical devices, which allow control on pressure and rates down-hole, to optimize production performance and ultimately improve oil reservoir recovery. The smart wells are installed with down-hole devices, that is, mechanical and electronic equipment that enables operators to control the wells remotely, without intervention using rigs or coiled tubing. A fundamental type of equipment used in smart wells is a down-hole mechanical valve called an interval control valve (ICV), which is preset with orifices with different hole sizes. The ICV is activated using an electronic pulse connected to an electrical cable embedded with a feedthrough packer. Fig. 7.1 shows the main components of a smart well.

During 2016, a total of 2200 wells had been completed with installations of flow-control devices (with and without sensing). Ajayi and Konopczynski (2008) compiled the value of smart wells from many fields around the world. They concluded that 9% of total oil recovery could be added by a single

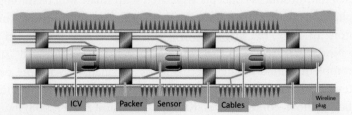

Fig. 7.1 Main components of a smart well.

smart well. They reported that up to 25% oil-recovery factor could be added by full field implementation of smart wells. The demonstrated economic value generated by smart wells includes the following.

- Saudi Aramco: A maximum reservoir contact project using multilateral wells in South Shaybah Field. The project features a multibranch well with a total of 12 km of drilled holes using five segments controlled with ICVs. The well produced 12,000 b/d when compared with a traditional horizontal well of 1 km producing 3000 b/d.
- Statoil: A subsea water alternating gas (WAG) project using 10 wells to inject gas and water, alternating mode, in the Snorre B Field. The water and gas breakthrough were delayed by 6 months, on average, per producer well, keeping production plateau for longer time than expected without smart wells. The ICVs were installed to control water injection; whereas gas injection was controlled by time.
- Kuwait Oil Company (KOC): Onshore stacked multilateral wells with an ICV per branch and 20 internal control devices (ICD) in the Minagish Field. The well had 5000 ft lateral section for each branch using an ICV port per lateral. The water cut was reduced from 75% to 25% in a mature water-flooded reservoir.

The main components of a smart well (Fig. 7.2) can be generalized as follows:

- Down-hole flow-control devices: this category can be grouped as a series of valves such as ICVs, which are remote-controlled mechanism with variable orifice size, and ICDs, which are preset mechanisms with fixed orifices.
- Down-hole sensors: electronic or mechanical devices that send signals to the transmitters.
- Transmitters: electronic devices that send different signals to the controllers.
- Isolation packers.

Fig. 7.2 Main components of a generic smart well, packer to isolate section, sensors, and interval control valve (ICV).

7.2 TYPES OF DOWN-HOLE VALVES

Service companies have created different types of control valves, which are classified based on their structure and main functions. The main categories include passive, autonomous passive, and reactive-actionable valves, which are all explained in this section.

7.2.1 Passive Valves

This classification includes ICDs, which restrict flow by creating additional flow rate-dependent pressure drops. The ICD generates an equalized pressure drop across the horizontal section avoiding early water- or gas-flow break-through. The ICD is preinstalled at the surface with the completion jewelry; once installed the orifices cannot be adjusted through time. It is called "passive" because there is no adjustment available on the orifice geometry. However, the potential benefit of an ICD is that this mechanism creates a homogenous pressure drop, equalizing the flowing bottom-hole pressure across the lateral length. Fig. 7.3 shows a schematic example of an oil well completion with and without ICDs which illustrates the reduction of gas or water coning by equalizing pressure distribution along the lateral.

7.2.2 Autonomous Passive Valves

A relatively new breakthrough technology developed by different service companies, the autonomous ICD (AICD) was created to bypass the water and gas by using centrifugation principle or differential density. This new tool is a self-regulating flow control device capable of controlling the fluids flowing through internal discs. Fig. 7.4 shows an example of an AICD, which shows oil and water flowing together through the ICD orifice; gravity and differential density cause oil to go directly to the outlet of the ICD,

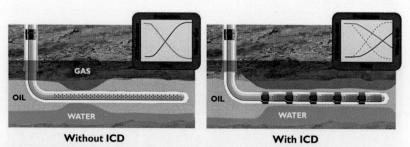

Without ICD   **With ICD**

Fig. 7.3 Comparison of water or gas front with or without ICD completion.

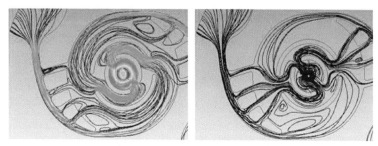

Fig. 7.4 Mathematical model representation of fluidic diode ICD (prototype of an AICD) with oil and water flowing through the internal orifice. Left figure shows water flowing at high velocity and high differential pressure. Right figure shows oil flowing at low velocity and low differential pressure. Note water flow takes larger pathway (inner circle) whereas oil takes shorter path. The experiment was carried out only single phase. *(Taken from Greci, S., Least, B., Tayloe, G., 2014. Testing results: erosion testing confirms the reliability of the fluidic diode type autonomous inflow control device. In: SPE-172077-MS. https://doi.org/10.2118/172077-MS, with permission.)*

whereas water goes around in each inner circle (longer path) and thus the mass flow rate of water is restricted. Greci et al. (2014) have carried out single-phase experiment by passing water or oil through internal orifice at different pressure drop showing changes in flow paths (Fig. 7.4).

7.2.3 Reactive-Actionable Valves

This category includes the ICV, which allows down-hole control of the well flow by turning a valve on or off, or by gradually controlling the choke size position. Generally, an ICV can be configured through function, actuation, and size. *ICV function*:

- Binary mode (on or off). Each segment of a well could be 100% closed or 100% open.
- Discrete multi-position: the most typical ICV. The device is set up with a movable collar, orifice (with different sizes), and static nozzle.
- High resolution with infinite variability: provides variable flow control (choking) with a customizable flow trim element. This flow trim element, along with pressure and temperature measurements, enables the calculation of accurate flow estimates. There are an infinite number of choke size positions between fully closed and its maximum open orifice position that can be chosen remotely by an operator.

ICV actuation:

- direct mechanical actuation (hydraulic balanced)
- hydraulic-spring return

- electric hydraulics
- mechanical override facility

ICV size:

- completion casing size for 6–1/2 in., 5–1/2 in., 4–1/2 in., and 3–1/2 in.
- hydraulic-spring return
- electric hydraulics
- mechanical override facility

7.3 SURFACE DATA ACQUISITION AND CONTROL

The down-hole equipment integrated by cables, connectors, gauges, and terminators are connected through the tubing to the control panel at the well surface. Fig. 7.5 shows the main configuration to acquire data and control ICV position for supervisory control and data acquisition (SCADA). The control panel receives pressure and temperature data in real time for the well segments completed with ICVs. Inside the control panel, the RTU sends the pressure and temperature signal to a SCADA system. Production engineers utilize a flow rate-based model and infer which segment is flowing, water, oil, or gas. Engineers decide which valve needs to be gradually controlled and send back a signal to the control panel to change the down-hole choke size.

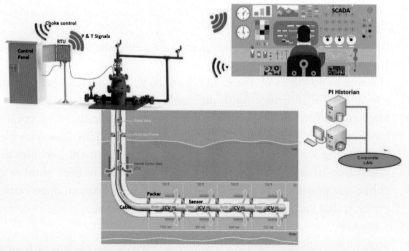

Fig. 7.5 Surface data acquisition and control using smart wells with ICV control.

7.4 SMART WELL APPLICATIONS

Control inflow of gas and water fluids: One of the main objectives of smart wells is to control the early water or gas breakthrough for primary or secondary recovery. Horizontal wells are ideal candidates because water or gas is mainly produced at the heel of well (due to the high-pressure drop). An ICV or ICD helps reduce the pressure drop or at least equalize the pressure across the lateral section. Water or gas can be controlled through time per segment in the lateral section.

Control commingling wells: Produce multiple zones to produce vertical wells. Smart wells enable exploitation of uneconomical production (one well per interval) and acceleration of reserves. It is a practice that is very well accepted by many national oil companies (NOC). The reserves per each reservoir can be drained using one common wellbore but controlled independently by using an ICV and packer per interval, as shown in Fig. 7.6. Any individual layer that has water or gas breakthrough can be shut off or controlled, which maintains overall oil cut.

Auto-gas lift injection: A gas interval zone in the upper section and oil in a lower section can easily be completed using smart wells. An ICV can be set up in the upper layers to control the necessary gas volume to lift the oil to the surface. Also, the water zone in the upper section can be reinjected to the lower section (with the oil) to reenergize the reservoir zone; this method is called a controlled water–dump flood. Fig. 7.7 (left) shows typical examples of auto-gas lift controlled at the surface.

EOR/IOR optimization: This is an area where smart wells could have a significant contribution to increase the oil-recovery factor. With an

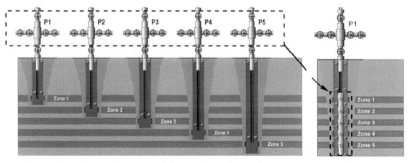

Fig. 7.6 Application of a smart well used to produce independent reservoir layers; the production from each layer is commingled and produced through a single well.

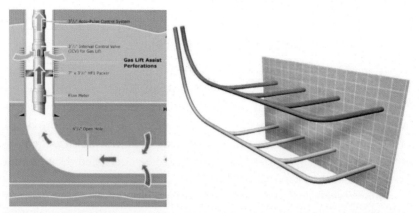

Fig. 7.7 (Left) Application of smart wells using the auto-gas lift injection; the gas in the upper section is controlled to lift oil through the tubing. (Right) An application of smart wells for an EOR process.

appropriate design, the ICVs are set up to control the contribution per segment in the lateral, particularly for those intervals with high permeability. The ICV orifice can be reduced or shutoff to avoid excessive water/gas production, or, using an optimized 3D reservoir model, to predict water breakthrough and then control the choke size to prevent the early water breakthrough without well intervention.

Another important application is using ICVs in well injectors. It has been demonstrated that smart wells can distribute the total water distribution to focus the injection in upswept zones. Fig. 7.7 (right) shows how an ICV can be used to maximize the oil displacement from the bottom to the top by injecting water to the bottom or gas to top using this technique.

Monitoring production in real time: Smart wells equipped with pressure and temperature sensors can analyze pressure losses, manage sandface pressure, and minimize the shut-in period for pressure buildup tests, and instantaneously capture data for unplanned well shut-in. As a result, smart wells help to reduce production downtime and accelerate production. Completion intervals with high water cut or GOR (unwanted) can be turned off temporarily or permanently.

7.5 SMART WELL PERFORMANCE

Usually, oil companies use production logging tools (PLTs) to evaluate and calibrate ICV valves. However, the cost of running a PLT in horizontal wells is so expensive that sometimes a virtual alternative is used—one

Table 7.1 Smart Well Experiment Showing Production Test Closing and Opening Three Valves per Experimental Design (Test No. Shaded Column Header)

Smart Well Test	Regular Test	Test 1	Test 2	Test 3	Test 4	Test 5	Test 6	Online 100% Open
Duration, h	24	4	4	4	4	4	4	24
Interval 1	On	On	On	Off	Off	Off	On	On
Interval 2	On	On	Off	On	Off	On	Off	On
Interval 3	On	Off	On	On	On	Off	Off	On
Oil STB/d Water Bbls/d Gas Mscf/d	Qo 1,500 Qw 600 Qg 2,000	Qo 1,250 Qw 300 Qg 1,680	Qo 800 Qw 300 Qg 1,066	Qo 600 Qw 500 Qg 1,366	Qo 350 Qw 320 Qg 480	Qo 820 Qw 380 Qg 1,020	Qo 560 Qw 150 Qg 728	Qo 1,450 Qw 640 Qg 1,780
Total Liquid STB/d	**2,100**	**1,550**	**1,100**	**1,040**	**670**	**1,200**	**710**	**2,090**

that requires implementing practical production tests using real-time data (such as pressure and temperature data) to evaluate dynamic wellbore and reservoir parameters (time-dependent properties), such as water saturation or pressure depletion. For these cases, we provide a technique to substitute a typical PLT with virtual PLT data, which is described in this section.

7.5.1 Production Test for Smart Wells

An experiment in an oil well in the Middle East with reservoir pressure almost constant (because of water injection) was used to diagnose the production influx per interval and detect offensive intervals in high water cut. The oil well produced 2100 STB/d of liquid, 1500 STB/d of oil, with a water cut of 28%. The water cut in the well suddenly increased from 10% to 28% in less than 30 days. The test was conducted with an experimental design of full-open (on) and full-closed (off) valve options on three intervals as described in Table 7.1 (where each column is a test protocol for the valves, shaded). The test duration per phase was 4 h per period (until pressure stabilized) for a total of 24 h.

Table 7.1 shows that interval 2 has the highest water rate with a water cut of 32% (380/1200, Test 4) and interval 1 has a 21% water cut. The test reveals that the intervals do not behave equally when production is commingled or producing alone: it is a case of "$1 + 1 \neq 2$". When the intervals produce in commingle mode, the flowing bottom-hole pressure is dominated by that interval with less pressure drop ($P_r - P_{wf}$). This test is frequently used to provide additional well performance analysis and evaluate matrix conductivity (kh), skin factor (S), and static bottom-hole pressure (P_e^*) in each segment. Also the test can be useful to calibrate a virtual PLT data.

7.5.2 Virtual PLT

Wireline PLTs are mechanical and electronical devices, such as turbines or spinners, which rotate at the fluid flow velocity. The PLT is used to register the fluid velocity and the fluid pressure and temperature per interval. The tool is calibrated using PVT data and down-hole tubular data to estimate the fluid flux per interval. The tools are frequently logged to measure the influx per interval and identify intervals with unwanted fluids or intervals with very low production targets (Fig. 7.8). Running a PLT in a horizontal section can be an expensive and risky operation. Smart wells are commonly equipped with pressure and temperature gauges at each ICV, which can reduce the number of well interventions needed to run PLTs. In this scenario, a steady-state hydraulic simulator can be used to evaluate the velocity, pressure drop, and production influx per interval and therefore identify the flow regime to create a virtual PLT based on a physical model. In this sense, the model should be set up with rock and PVT properties across the lateral

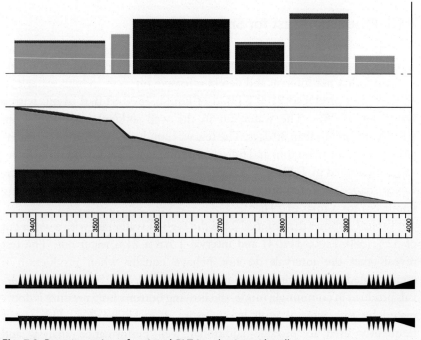

Fig. 7.8 Representation of a virtual PLT in a horizontal well.

section. More importantly, the hydraulic simulator should be capable of computing the effect of the ICV by using a valve flow coefficient (Cv) that has been measured in lab tests for multi- and single-phase flow.

To simulate and reproduce physical models that interact with smart well completions, an automated model-calibration workflow that optimizes down-hole valve settings for maximized oil-recovery factor can be developed. In short, such an automated workflow enables reactive and proactive decisions to control unwanted production (water/gas, solids, fines, etc.) from the smart well. The objective of this workflow is to allow the data exchange and iteration between subsurface and surface models—such as the reservoir simulator, hydraulic completions, and nodal wellbore and network analysis—and to periodically update the regional/stor reservoir model associated with the drainage area of the smart well completion. Moreover, the workflow is required to model a semi-analytical wellbore model to perform the optimization of vertical lift performance (VLP). Such a workflow needs to facilitate the following engineering functionalities:

- Receive, update, and allocate the real-time production data from the remote-controlled system.
- Perform multilevel matching and calibration of a history-matched simulation model with observed pressures and rates at the smart well completion interval, well tubing, and surface production.
- Execute reactive control by optimizing the down-hole valve setting in response to local data and predicted short-term fluids behavior.
- Execute proactive control by optimizing the down-hole valve settings at multiple intervals to maximize oil recovery using both local nodal properties and simulated predictions.

An example of an automated workflow for smart well calibration is given in Fig. 7.9.

Dynamic calibration of a simulation model with observed pressures and rates at the smart well completion level (i.e., per individual segment of the ICV) is enabled through design, integration, and reconciliation of a virtual PLT profile, which represents a probability distribution of a parameterized reservoir property (such as permeability, water saturation, etc.). Such a distribution is derived as discrete conditional probability of reservoir property (m) given the fluid flow rate (f) per perforation interval (i) of the ICV segment (s) of a smart well (w) as expressed in Eq. (7.1):

$$\widetilde{m}^w_{s,i|f} = \mathrm{Prob}\left(M = \hat{m}^w_s \mid F = f\right) = \frac{Prob\left(F = f \cap M = \hat{m}^w_s\right)}{Prob(F = f)} \qquad (7.1)$$

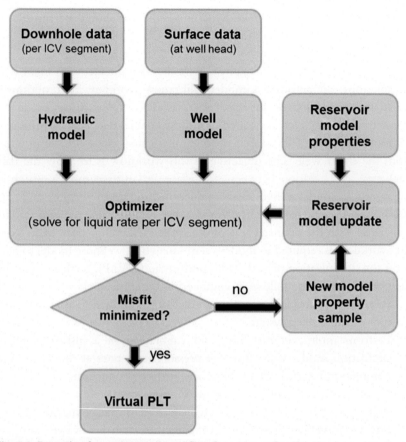

Fig. 7.9 Example of an automated workflow for smart well calibration by generating a virtual PLT profile.

where M and F correspond to overall sampling domains of the reservoir model property and fluid flow, respectively. The $\hat{m}$ symbol corresponds to the average (mean) value of model property m in Eq. (7.2). The sample from such conditional distribution, corresponding to "*New Model Property Sample*" in Fig. 7.9 becomes proportional to the fluid flow density in the ICV segment:

$$\widetilde{m}_{s,i}^{w} \propto \frac{\sum_{i,s} (f_i \times \Delta i)}{F \times I} \tag{7.2}$$

where Δi corresponds to the width of perforation interval i and I is the total width of perforations per ICV segment. An optimization algorithm iteratively samples $\widetilde{m}_{s,i}^{w}$ and minimizes the misfit between the dynamic smart well response and the observed production profile (e.g., water cut, pressure, GOR) that renders, for example, the modeled liquid rate profile that matches the well completion profile from the ICV segmentation. By further calculating the water flow rates per ICV segment, any type of water flooding optimization can be facilitated. To quantify the uncertainty in the smart well model response (i.e., the virtual PLT profile), the optimization method can further be integrated into multiple stochastic realizations of reservoir properties.

7.6 SMART WELL MODELING AND CONTROL

For wellbore modeling, the oil and gas well deliverability per segment can be calculated using inflow and outflow well performance analysis. Most of the commercial simulators can reproduce the effect of an ICV in a wellbore. Konopczynski and Ajayi (2004) have described the fundamental concepts to evaluate the well performance for single- or multiple-zone reservoirs using ICVs. The IPR and Vogel expressions can be easily used under the same assumptions that are used under radial and Darcy's flow conditions. We show a methodology to estimate ICV performance as follows:

- Estimate maximum well rate potential from an IPR/Vogel model.
- Productivity index, $PI = Q/(P_{res} - P_{wf})$.
- Calculate the pressure drop across the choke ($P_{out} - P_{in}$) required delivering the expected rate reduction/increase.
- Determine the flow trim characteristic by lab testing the flow coefficient at each choke position of pressure drop.
- Carry out in the lab the mechanical design of the flow trim for the desired well behavior.

The Perkins model (1993) is the most applicable equation to predict the subcritical pressure drop across a valve using Eq. (7.3):

$$P_{out} - P_{in} = \frac{-Q \times |Q| \times \rho}{Cv \times Cv \times \rho_{wstd}} = C \times \gamma_{mix} \times Q_{tot} \sqrt{\frac{1}{Cv}} \qquad (7.3)$$

where
P_{out} is the flowing BHP at the outlet (psi) of the ICV,
P_{in} is the flowing BHP at the inlet of the valve (psi),
C is the conversion factor,
Q is the total fluids in USG/min,

ρ is the fluid density of the mixture phases (lbm/ft^3),

Cv is the flow coefficient in USgal/(min psi$^{0.5}$) (which is measured in the lab of the service provider),

ρ_{wstd} is the water density measured at standard condition in lbm/ft^3, and

γ is the liquid gradient of the mixed fluids.

The Cv performance is represented by a choke valve opening versus flow rate table determined at lab condition using compressible fluid at high temperature and pressure. Fig. 7.10 shows a pressure drop versus oil flow rate delivered by an ICV flowing oil fluid at 2600 psi and 200°F in tubing size of 3–1/2 in., oil API of 30 and gas gravity of 0.81. Cv values are observed in each position (Pos).

7.6.1 Single-Zone Control Analysis Using an ICV

The results of the Perkins equation at different pressure drops across the valve were superimposed on the IPR and VLP plot showed in Fig. 7.11. The values of the pressure differentials may be considered at discrete flow rates as a fraction of the maximum flow rate, such as 20%, 40%, 60%, etc., and the operating points at the perforation node are represented by the corresponding points labeled A, B, C, D, E, and F. The number of discrete positions corresponds to set positions of the flow control valve. Konopczynski and Ajayi (2004) referred to these as attenuated IPR curves.

7.6.2 Multiple-Zone Control Analysis Using ICVs

The process for multiple-zone intervals is very similar to the process for a single zone. However, the main limitation is that the process assumes that all zones have the same reservoir pressure, fluid composition, and bubble point. Another important assumption is that the Cv flow coefficient and devices are equal for all zones. They established two methods to estimate the pressure across the valve:

- *Constant pressure.* This method assumes that only the flow from Zone A is modulated while the production from all other zones is maintained at a constant rate. To satisfy this constraint, the flowing bottom–hole pressure must be held at a constant value.
- *Independent flow.* This method assumes that the fluids flow only from the zone of interest through tubing and the other zones are shut-in; it is similar to the single-zone method.

For this particular example, Konopczynski and Ajayi (2004) used a tubing size of 5–1/2 in. to independently flow in five zones, with IPs that vary between 3.4 and 7 b/d/psi at a surface wellhead pressure of 150 psi. Fig. 7.12 (left)

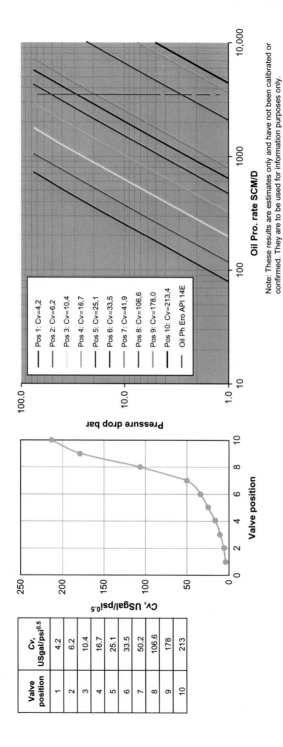

Fig. 7.10 Typical values of Cv coefficients for a tubing size of 3–1/2 in., API fluid of 30, 200°F and 2600 psi flowing BHP. *(Taken with permission from SPE 90664.)*

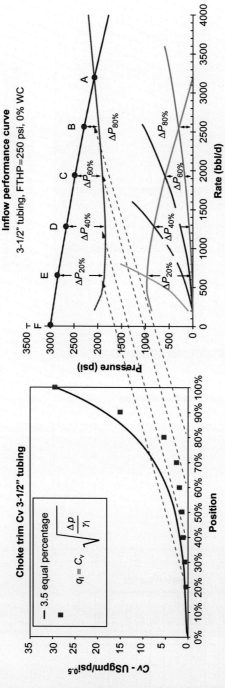

Fig. 7.11 IPR and VLP curve for a smart well with a single-zone interval. *(Taken with permission from SPE 90664.)*

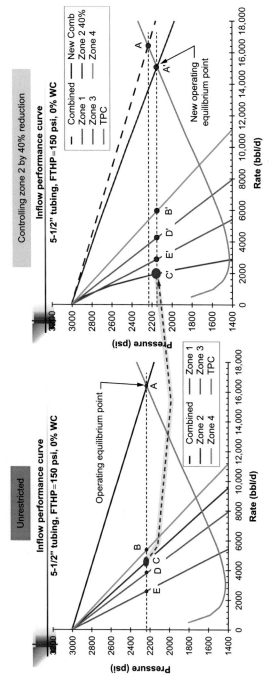

Fig. 7.12 IPR and VLP curves for a smart well with multiple-zone intervals. *(Taken with permission from SPE 90664.)*

shows the IPR curves for individual zone and the combined IPR for commingled unrestricted production. If the operating objective is to maintain constant flow at Zones 1, 3, and 4, and to restrict flow in Zone 2% to 40% of unconstrained flow, then the flowing bottom-hole pressure must be restored to the original pressure. Controlling the valve in Zone 2, the attenuated IPR curve for Zone 2, with a flow control valve setting of 40%, is shown in Fig. 7.12, along with the unrestricted IPR curves of the other zones.

7.6.3 Coupling Wellbore and Gridded Simulators to Model ICVs

Wells with ICVs can be modeled using commercial well performance software coupled with a 3D numerical model. The numerical model simulators use a finite difference three-phase simulator, which has been described widely (Coats et al., 2004; Shiralkar et al., 2005). The simulator has a gridded-cell wellbore and network tubular models that are connected to "wellnodes" and linked by "connection." Valve connections are between well nodes for tubing and annulus sections. Fig. 7.13 shows a smart well

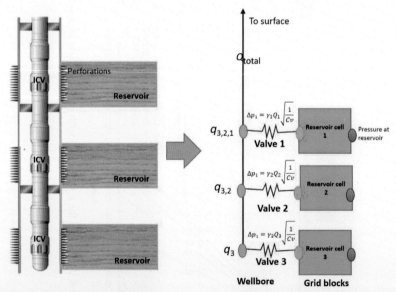

Fig. 7.13 Schematic representation of a smart well showing wellbore nodes connected to reservoir cells. (A) Mechanical configuration for three interval control valves. (B) Schematic for representing the mechanical configuration in a reservoir or well simulator using wellbore nodes coupled to reservoir cells.

schematic with a representation of coupling wellbore nodes with reservoir cells. It is recommended to set up an ICV valve per reservoir cell; in those scenarios where the cell is bigger than the segment where the ICV is completed, it is suggested to create a local grid refinement (LGR) to connect one cell with one segment.

The Perkins equation (7.3) is coupled with the numerical models with a surface-network capability assuming that these parameters are known: molar rates of hydrocarbon components, water rate at stock tank conditions, temperature at the outlet of the valve (used in flash calculations), valve control, and the flow valve coefficient profile. The simulator solves the pressure equation across the valve with the following procedure solution algorithms:

- Determine the number of hydrocarbon phases and the compositions and compressibility factors of each phase using the phase-equilibrium calculation.
- Determine the densities of oil, gas, and water phases (flash calculations).
- Determine the specified type of mass rate.
- Calculate density of the fluid mixture.
- Compute interpolation values of valve coefficient as described in Section 7.6.
- Determine the pressure differential across the valve.

Fig. 7.13 shows a schematic representation of how the fluids flow through the ICV, the reservoir produces through perforated casing to the annulus, and valves control flow from the annulus into the tubing contained inside the casing. Annular flow in the casing is blocked by packers and between three perforated sections of the wellbore.

The VLP of the wells could be represented using hydraulic tables from nodal analysis or any industry-standard analytic multiphase flow correlation (e.g., Beggs and Brill, Hagedorn and Brown, etc.), which can be applied to the connection between ICV nodes and the vertical section of the wellbore. The VLP is required to complete the surface calculation of production rate and pressure, and to estimate the pressure drop across the vertical section.

7.6.4 Modeling ICDs for Oil Wells

The ICDs are designed to control the down-hole flow distribution by creating a frictional pressure drop across the orifice; it is the ICD that acts as setting up several chokes in front of the casing. This device is created to reduce the pressure drop at the heel of a horizontal well or reduce the pressure drop in front of a high-permeability segment in the horizontal section, equalizing the pressure drop along the horizontal section. As a result of this reduction in pressure drop, the ICDs can reduce the early water or gas breakthrough (preferably at the heel) and thus increase the oil-recovery factor. There are different types of ICDs, for example, channel, nozzle, orifice, helix tube, and hybrid. The most common types applied in the oil field are the nozzle and helix.

Henriksen et al. (2006) and later on Zhu and Hill (2008) have demonstrated the importance of ICDs in high-permeability reservoirs and also demonstrated that the main control property is the pressure-friction losses generated across the ICD valve. To model the subcritical pressure drop across an ICD, the following expressions (Eqs. 7.4, 7.5) can be used:

$$P_{out} - P_{in} = C \times \frac{\rho_{mix}}{2} \times \left[\frac{q}{A \times Cv} \right]^2 \qquad (7.4)$$

$$\rho_{mix} = \text{water.vol} \times \rho_w + \text{oil.vol} \times \rho_o + \text{gas.vol} \times \rho_g \qquad (7.5)$$

where
C is the conversion factor,
P_{out} is the flowing BHP at the outlet in psi,
P_{in} is the flowing BHP at the inlet of the ICD in psi,
q is the total flow rate of fluids in ft^3/s,
ρ_{mix} is the fluid density of the mixture phases at pressure and temperature in lbm/ft^3,
Cv is the flow coefficient in $USgal/(\min psi^{0.5})$ (which is measured in the lab of service provider),
ρ_o, ρ_w, and ρ_g are the oil, water, and gas densities at pressure and temperature in lbm/ft^3,
water.vol, oil.vol, and gas.vol are the volume fractions of the phases in the wellbore at time t, and
A is the cross-sectional area of the ICD in ft^2.

Henriksen et al. (2005) and Birchenko (2010) have demonstrated in their field and simulation studies, respectively, that the permeability distribution along the horizontal trajectory is the most important parameter to design ICD types, and determine the number of ICDs required along the horizontal wellbore and the number of packers required. By generating a pressure drop by using appropriate packer spacing with ICDs, the flow velocity is reduced, which results in longer sustained oil production before water and gas breakthrough to the wellbore. They concluded that ICDs could delay the water breakthrough compared with an open-hole completion. Daneshy et al. (2010) and Twerda et al. (2011) have conducted research on ICDs and determined that they are excellent tools to control water coning in water flooding. In fact, their research based on simulation study found that ICD use leads to an increased oil-recovery factor, but not necessarily to increased net present value (NPV) or internal rate of return (IRR). They also found that ICDs do not work properly in gas coning or low permeability reservoirs for their cases.

Carvajal et al. (2013a,b) have shown the impact of ICD completions coupled with a 3D gridded, three-phase flow numerical model applied in a horizontal well with permeability variation among 1, 100, 10, 1, and 100 md. They simulated one horizontal well producer of 3000 ft with 40 ICDs segmented into five regions, isolated with swell packers. The objective was to reduce the water cut and delay the water breakthrough from an active bottom aquifer. The simulation results are shown in Fig. 7.14, which compare an open-hole completion with an ICD completion.

The results show that water breakthrough occurred earlier in the open-hole completion at 500th day; whereas, with an ICD completion, the water breakthrough occurred at 630th day. Moreover, the waterfront using the

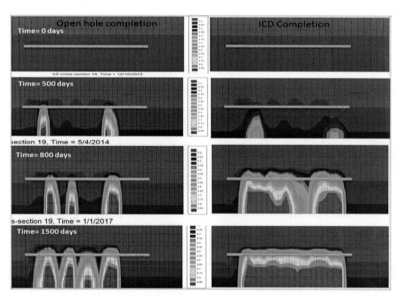

Fig. 7.14 Vertical cross section (*XZ*) along a west lateral section of a multilateral well showing the water saturation front at different time steps for an open-hole completion (left) and ICD completion (right). *(Taken with permission from SPE 164815.)*

ICD is equalized along the horizontal section, which results in an excellent oil-sweep efficacy from the oil–water contact (OWC) to the wellbore.

7.6.5 Modeling AICDs for Oil Wells

As mentioned in Section 7.2.2, AICD is the newest family of ICDs that are self-regulating fluidic diodes capable of restricting the flow of fluids based mainly on differential densities and viscosities. Contrasting with an ICD, which generates high-flow restriction for heavy oil, the AICD can sometimes be more feasible in restricting water and gas than a traditional ICD, which is attributed mainly to the differential density. One of the main benefits of AICDs is that it can generate spontaneous changes in fluid flow restriction without control lines, moving parts, or download mechanical or electrical devices for well intervention and moving choke size.

In some laboratory experiments with single-phase flow (oil or water), Least et al. (2012, 2013) showed for the first time the performance of AICDs using oil fluids from 3 to 200 cPs. They found that the pressure drop across an AICD valve could be governed by the ratio of fluid density and viscosity compared with calibrated density and viscosity of the fluid tested at the lab. To model the subcritical pressure drop across an AICD, the following equations can be used:

$$P_{out} - P_{in} = X \times \frac{\rho_{mix}^2}{\rho_{std}} \times \left[\frac{\mu_{std}}{\mu_{mix}}\right]^{\beta} \times q^{\alpha} \qquad (7.6)$$

$$\rho_{mix} = \text{water.vol} \times \rho_w + \text{oil.vol} \times \rho_o + \text{gas.vol} \times \rho_g \qquad (7.7)$$

where
P_{out} is the flowing BHP at the outlet in psi,
P_{in} is the flowing BHP at the inlet of the valve in psi,
q is the total flow rate of fluids in ft^3/s,
ρ_{mix} is the fluid density of the mixture phases at pressure and temperature in lbm/ft^3,
ρ_{std} is the fluid density measured at laboratory condition used for calibration proposes in lbm/ft^3,
ρ_o, ρ_w, ρ_g is the oil, water, and gas densities at pressure and temperature in lbm/ft^3,
μ_{mix} is the mixture fluid viscosity calculated at the average of $(P_{out} - P_{in})$ in cP, μ_{std} is the fluid viscosity measured at laboratory condition used for calibration proposes in cP,

μ_o, μ_w, and μ_g are the oil, water, and gas viscosities at pressure and temperature in cP,

water.vol, oil.vol, and gas.vol are the volume fractions of the phases in the wellbore at time t,

X is the device strength parameter, which is measured during the calibration process in psi/(lb/ft^3 ft^3),

β is the viscosity exponent measured during the calibration in the lab (dimensionless), and

α is the rate-dependent exponent measured during the calibration in the lab (dimensionless).

Carvajal and Torres (2014) have shown the impact of an AICD completion coupled with a 3D-gridded, three-phase flow numerical model applied in a horizontal well with three different oil viscosities: 1, 10, and 100 cP. They found that an AICD does not have a significant impact on the production for those cases with oil viscosity below 1 cP or viscosity close to water viscosity. However, they found that in a scenario with a viscosity ratio two times more than water viscosity, the AICD had a significant impact on oil recovery. They simulated one horizontal well producer of 5000 ft with an injector horizontal well in a reservoir of 10 cP oil viscosity and permeability variation between 1 and 100 md. The model was set up with 40 AICDs (a grid block per AICD, grid block size = 20 ft). The simulation results are shown in Fig. 7.15 comparing an open-hole completion with an AICD and a full producer completion with an AICD.

The producer well and the injector are separated by a distance of 3 km. The water breaks through the producer at 380 days after the first water injection. In both cases, the water breakthrough time is the same; water arrived at the middle of the lateral. Therefore, the AICD does not contribute significantly in delaying the water breakthrough. However, before day 1500 of production, using streamline simulation, the water is deviated and bypassed the wellbore, whereas oil was directly injected into the wellbore, following exactly the same physics principles of kinematic differences between fluids $(\rho_o \cdot \mu_w / \rho_w \cdot \mu_o)$.

Finally, they found that using AICDs, the oil-recovery factor was increased by 20%, and water is significantly reduced from 60% to 20%, compared with an open-hole completion. More than 20% increment was observed after water breakthrough. The AICD had a significant impact on controlling water after water breakthrough. Nevertheless, we believe that the best combination is to set up both an ICD and an AICD, which could reduce the water breakthrough time.

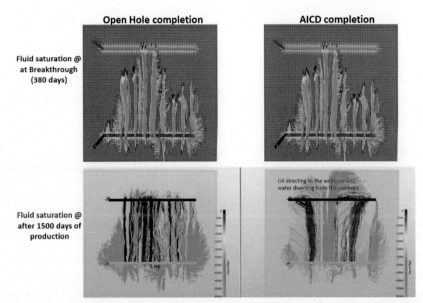

Fig. 7.15 Aerial section of a 3D numerical model showing a pair of horizontal wells (producer-injector). (Left) Injecting water without an AICD and (right) 40 AICDs plus packers. *(From Carvajal, G., Torres, M., 2014. Coupling AICD to a wellbore and numerical model simulation. In: Presentation About AICD Progress. Presented at the Novel Techniques for Reservoir Management, SPE, Frontier of Technology—Reservoir Technology Forum Held in New Mexico, USA, November 4–9.)*

7.7 OPTIMIZING FIELD PRODUCTION WITH SMART WELLS

Optimizing smart wells is practically a new science that requires a series of engineering tools (software and hardware) to evaluate and compute the effect of valve settings over production. The best way to optimize production performance with a smart well is to couple a wellbore hydraulic well performance simulator with a 3D gridded numerical model. Ajayi and Konopczynski (2003), Elmsallati and Davies (2005), and Nikolaou et al. (2006) are among the first references found to predict and optimize ICV performance in a single well, compared with conventional completions.

Saputelli et al. (2009) have presented the first surface-well model connected to a full field numerical simulator; this paper describes a process to optimize long-term economic return in oil reservoirs with water flooding by optimizing the number of wells, usability of ICV valves, and operating schedules. They found that by systematically optimizing the ICV setting in injector and producer wells, the ultimate oil recovery was improved 3% by reducing water injection by 8%, and economic value (NPV) was improved 2.3 times, compared with conventional completions.

7.7.1 Control Modes

The "best way" to optimize smart wells depends strongly on a company's philosophy of reservoir management. When controlling the choke size, the oil industry refers to this as reactive and proactive operations (Jansen, 2001; Kharghoria et al., 2002). With increasing use of 4D seismic, passive control is also another type of choke control. These three control modes are summarized and distinguished below.

- *Reactive control.* If the GOR or water cut (wc%) exceeds its target maximum allowable production during a period of time, the ICVs are controlled to reduce GOR or wc%, or at least to prevent further increase in wc% until the target is reached. The philosophy is based purely on observation and reaction; if something happens, then the ICVs are changed. It is the most frequently used control method. However, because of the restriction, this control often results in poor oil-sweep efficiency, and in certain regions of the reservoir, the oil could be bypassed and not drained well (Essen et al., 2010). In this control, the essential toolkits or software may be well surveillance, production monitoring, and nodal analysis.
- *Proactive control.* On the basis of the prediction of the GOR or wc% production profile, the value of ICV settings are anticipated and set up before the water or gas breakthrough occurs at the well. It is believed that the gas or water flooding (either from injection or gas cap/aquifer) is still away from the wellbore at some distance between the wellbore and the reservoir. The philosophy is based on trusted reservoir data or a physics-based reservoir model (i.e., 3D numerical simulator or 1D analytical application). If the reservoir is very well characterized in terms of fluid-rock properties and the 3D numerical model can capture the main heterogeneities, then proactive control can maximize the oil-sweep efficiency. However, this control is sometimes found to be impractical compared with reactive control, because of the time it takes to run the batch process and integrating geological and reservoir data.
- *Passive control.* 4D time-lapsed seismic processes a seismic signal to interpret the signal attributes and to capture water movement during water

injection at different time steps (preferably every 4 months). This capability means that the waterfront location between wells can now be estimated and thus water breakthrough can be delayed with high accuracy, compared with proactive control. The ICV settings can be changed depending on the waterfront movement. For example, when the waterfront is approaching the wellbore, the ICV in the segment can be completely off, which maximizes the oil-sweep efficiency. The philosophy is based on trusted 4D seismic processing and interpretation, and how the 3D numerical model captures the seismic data. Passive control is the most sophisticated control because it can be used at the right time with the right tool. However, the use of 4D seismic and this control is very expensive and requires extensive seismic data and interpretation.

Cullick and Sukkestad (2010) have modeled smart wells by applying ICVs in complex well architectures including multilaterals wells. They compared the production performance by using a well completion without an ICV (open-hole completion), reactive control (fixed policy), and proactive control (optimized policy). Fig. 7.16 shows the results of different control modes. The fixed policy is an optimization technique that runs the simulation model and checks water cut value in each simulation's time step against the threshold. If the threshold is met or exceeded (e.g., 80% of water cut), the valve setting is reduced by one unit. The process is continued until the ICV is fully closed. With the optimized policy, a set of threshold ranges and increment sets are provided and they are used as optimizer to control the 3D numerical model. The optimizer seeks a combination of triggers and increments that maximizes the oil-recovery factor while it reduces both water production and injection. Each valve has its own best threshold trigger water cut and

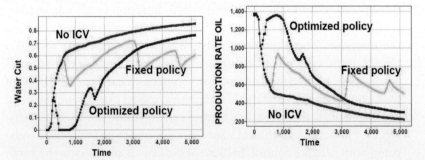

Fig. 7.16 Production profiles for no ICV, reactive (fixed policy), and proactive (optimized policy) controls in a horizontal well with three ICVs. *(Taken with permission from SPE 126246.)*

increment range, which generates a different setting history for each ICV. The results are shown in Fig. 7.16, where you can clearly see that using an optimized policy, the oil-recovery factor yields the highest oil production. The authors have reported that oil increased by 67% and water decreased by 47%, compared with no ICV completion.

In the Sabriyah-Mauddud Formation in North Kuwait, a horizontal smart well with five ICVs was installed in the field. Using the same approach presented by Cullick and Sukkestad (2010) and Carvajal et al. (2013a,b) showed a procedure to capture the main heterogeneity of the reservoir model, such as permeability high street. They created an automated process to couple real-time data, the reservoir model, and the surface system. The flow coefficient for 10 opening apertures of ICVs were set up in the simulator, and the optimizer changed the valve settings while the oil-recovery factor was maximized under the condition of maximum reservoir voidage replacement of 80%.

The optimizer coupled with the numerical model ran more than 100 possible combinations. When the objective function was reached, a global optimum point was determined as the best solution. They found that the best combination of ICV settings, year by year, is as shown in Fig. 7.17. The optimizer started with 100% open (position 10 out of 10), but noted that the valve position does not change monotonically; the valves open and close until the optimizer satisfies the best solution for the objective function. For example, valves 1, 3, 4, and 5 changes every year, but valve 2 does not change periodically. This difference is because of the high permeability street found in segments 1, 3, 4, and 5. The study concluded that by converting conventional wells to smart wells in the entire field in this reservoir under water injection, the oil-recovery factor is increased by 52% (increment) compared with conventional completions. Moreover, this approach stabilized the oil plateau for more than 5 years, while the water cut was going up and down with the ICV settings.

7.8 SMART IMPROVED OIL RECOVERY/ENHANCED OIL RECOVERY MANAGEMENT

Improved oil recovery (IOR) is a technical process that injects natural gas or water into the reservoir to increase oil reserves beyond the primary recovery or natural forces of the reservoir. Enhanced oil recovery (EOR) is defined by the US Department of Energy as a series of techniques using special fluids to increase the oil-recovery factor beyond the IOR process;

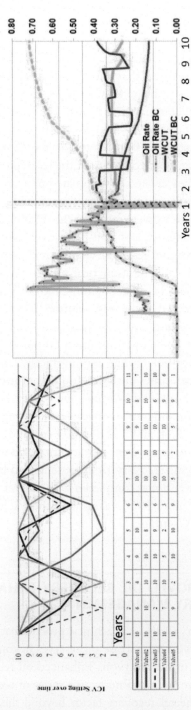

Fig. 7.17 Production profiles without ICV and proactive controls in a horizontal well with five ICVs (from Carvajal et al. 2013a,b). The plot compared a base case (BC) without ICV for both oil rate and water cut profiles with oil rate and water cut with ICV. The base case (without ICV) water cut progressively increases with time while oil is declined. Although using ICV, the water cut is controlled between 0.3 and 0.4, the oil rate production is almost flat. *(Taken with permission from SPE 164814.)*

regularly the range of improvement is between 30% and 60%. IOR/EOR has are divided three main categories: thermal fluids (steams and in situ combustion), chemical fluids (alkaline, polymers, and surfactants), and gas injection (CO_2, N_2, and fuel gas).

Use of EOR/IOR methods began in the 1950s, but the 1970s saw a dramatic increase in IOR implementation. During the 1980s and 1990s, oil companies used cased-hole logs, PLTs, pressure transient tests (PTT), and down-hole pressure gauges to evaluate the water and gas injection fronts. All of these techniques monitor production response, well to well (point to point), and are logged in the well sporadically over time (once per year) but are unable to predict water or gas breakthrough with accuracy. Therefore, the control mode was generally reactive and rarely proactive. The volumes of data managed by these tools do not exceed the kilobytes/days.

Modern EOR/IOR processes include a series of technology, hardware, and software specially designed for this operation, which allows real-time data capture of massive volumes of data (e.g., megabyte of data per minute). Using fiber optic cable in horizontal wells, temperature, pressure, and strain can be monitored and logged at every meter, detecting with high precision and on time, the segment of the well that is being invaded with unwanted or injected fluids. Some examples of these technologies include the following:

- Real-time use of a series of chemical, thermal, and ultrasonic sensors to monitor fluid chemical properties, such as liquid pH, chlorides, solids, minerals, ions, cations, temperature, stream quality (x%), wax, asphaltene, proppant, sand, tracers, etc.
- Down-hole fiber optics. These technologies include: (1) distributed temperature sensing (DTS), which captures several megabytes/day to register fluid temperature behind the casing along the horizontal and has been recently applied to monitor hydraulic fracturing and (2) distribute acoustic sensing (DAS), which collects terabytes/day to register strain sensing along the horizontal section.
- 4D time-lapsed seismic, micro-seismic, cross-well seismic, and vertical seismic profiling.

The combination of ICVs/ICDs with any of the above technologies has generated unprecedented information and analysis that was previously impossible to decipher because of data-processing technology limitations (Regtien, 2010; Clark et al., 2010). Smart IOR/EOR could be defined as a process that uses a series of smart components and automated functionalities, which cover the full DOF chain, including remote sensing, data acquisition, workflow automation, visualization, and collaboration (where decisions are made for ICV control). The main difference between smart and traditional IOR/EOR processes is that smart EOR uses:

- Massive subsurface (fiber optic or seismic) data streaming to monitor the frontal advance of injected fluids (water or steam).
- ICVs to control the well at down-hole condition, which allows better oil-sweep efficiency by focusing water, steam, or gas in the upswept/bypassed reservoir segments.
- Coupled surface, wellbore, and 3D reservoir models in simulators to generate scenarios to prevent early water, steam, or gas breakthrough to the producer wells.

7.8.1 WAG Injection Process

The WAG process is designed to improve sweep efficiency in order to reduce residual oil saturation after conventional water or gas injection and to control early water or gas breakthrough to producer wells. Depending on the fluid and rock types, viscosity, and wettability, water is injected into the reservoir for 2–6 months, followed by gas, and the cycle is repeated. Simultaneous water and gas (SWAG) injection is a variation where water and gas are injected simultaneously through the same tubing. Fig. 7.18 shows a traditional WAG process: water is injected into the reservoir followed by a slug of gas, and the process is repeated until water cut or GOR exceeds the economic limits.

For horizontal wells with lateral sections longer than 3000 ft, controlling the injection point is difficult due to the Toe-Heel Effect, which refers to most injection water going into the first ~1000 ft of the lateral, leaving the rest of the lateral with limited to no injection. Operators sometimes use down-hole control valves such as ICDs or ICVs to distribute the injection flow across the lateral section.

7.8.1.1 WAG Process With ICV

Carvajal et al. (2015) have proposed a continuous injection of water and gas slugs, injecting water through production casing and gas through tubing

Fig. 7.18 Traditional WAG process using vertical wells.

while selecting the optimum injection points for water and gas individually along the well lateral. ICVs and a new mechanical well configuration are used to enable this continuous injection. The process is called WAGCV. They set up a numerical model with the new ICV design by injecting gas and water at different locations in the lateral. These results show that the proposed process should improve oil recovery significantly compared to traditional WAG, because:

- Residual oil saturation is significantly reduced in all regions due to more homogenous oil sweep.
- Water and/or gas breakthrough is substantially delayed.
- As a result, oil-recovery factor increases more than 5% over traditional WAG.

The well uses two strings, where water is injected through casing and gas through tubing. The well uses a unique combination of down-hole valves, sleeves, packers, and fiber optic equipment. Down-hole valves and sleeves are automatically activated on or off to coordinate the multi-injection points. The times to activate specific injection points are defined by the optimizer software, which estimates fluid injection volumes (for gas, water, or both) and slug locations in the reservoir over time.

For illustrative purposes, a reservoir with high heterogeneity in permeability between 10 and 150 md (Fig. 7.19) was set up in a 3D numerical simulator that predicts the production profiles of water, gas, and oil. Horizontal wells, an injector and a producer 3000 ft apart, are configured in the model, each well with 4000 ft laterals.

The process modeled is described as follows:

A. Water is injected into the reservoir for a long period of time (Fig. 7.19A). Immediately water starts channeling into the high permeability regions, that is, Regions 1 and 3. All valves are 100% open.

B. After several numerical simulation iterations, the optimizer determines that a slug of gas should be injected through tubing into Region 1 with a specified slug size and specified daily rate (Fig. 7.19B). The gas slug is injected through tubing into Region 1, while water is injected across the rest of the lateral section. The sleeve in Region 1 is shut off, allowing gas injection through the tubing and blocking water injection in Region 1. Water continues to be injected into the rest of the lateral section.

C. After a period of time, the optimizer determines that additional slugs of gas should be injected through tubing into Regions 1, 3, and 5 at specific volumes and injection rates as shown in Fig. 7.19C. The gas slugs are

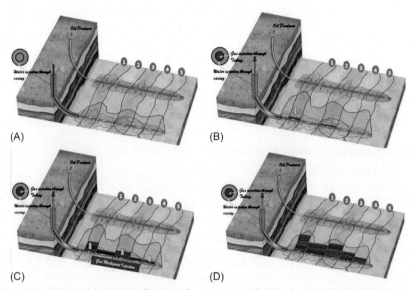

Fig. 7.19 (A) Initial injection of a slug of water controlled by the ICVs. (B) Initial injection of a slug of gas in Region 1 while water is injected across the lateral section. (C) Multipoint injection commences in Regions 1, 3, and 5, while water is continuously injected into Regions 2 and 4. (D) The workflow process showing the new properties. *(Taken with permission from EAGE white paper 2214-4609.)*

injected through tubing into Regions 1, 3, and 5, while water continues to be injected into Regions 2 and 4. Sleeves in Region 1, 3, and 5 are shut off to allow gas to be injected through tubing and block water injection into these regions.

D. The process is repeated in multiple cycles (Fig. 7.19D) to maximize the oil-recovery factor and minimize either wc% or GOR. Part of the objective is to reduce the residual oil saturation (Sor) after water injection into each region, delay early water or gas breakthrough, and achieve homogenous oil drainage across the reservoir. The process ends once wc% or GOR reaches the maximum production limit established by the operator.

7.8.1.2 WAGCV Numerical Simulation

The 3D reservoir simulation model was built with 1.4 million cells; it is a high-resolution reservoir model that covers a pilot area of one producer and one injector well, both horizontal. A black-oil fluid model was used with viscosity and oil density variation from the crest to the flanks of the geological structure. The reservoir sector was segmented into five regions.

For comparison with WAGCV, three processes that can help to increase reservoir recovery beyond primary depletion were selected: gas injection, water flooding, and traditional WAG.

In a separate exercise, the gas and water injection rates were optimized to determine the best rate of injection for gas and water, while the traditional WAG was optimized by changing the slug ratio with time. The WAG exercise showed that the best combination is 6 months of injecting water at 10,000 STB/D, followed by 6 months of injecting gas at 10 MMscf/D. In WAGCV, each region was set up with an ICV that controlled the injection rate. Oil rate and cumulative oil versus time are displayed in Fig. 7.20.

The oil rate profile for the gas injection scenario (purple line) shows a moderate plateau of 4500 STB/D for almost 2 years, with a total cumulative oil of 12.5 million bbl in 15 years. The oil rate profile for the water injection scenario (blue line) exhibits the longest plateau, but the decline rate is sharp compared with the other scenarios, so the cumulative oil by water injection is approximately the same as for gas injection: 12.2 million bbl. The traditional WAG process follows the water injection profile, but production goes up and down (red dotted line) at a controlled decline rate for 10 years, after which production drops sharply. The total cumulative oil is 16 million bbl, 3.5 million bbl more than classic injection whereas WAGCV technique (green line) offers a better production profile, maintaining a slow decline rate for almost 10 years. The total cumulative oil improves to 18.0 million bbl,

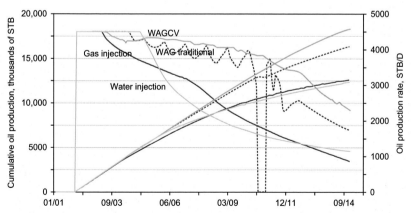

Fig. 7.20 WAGCV simulation: results of oil rate and cumulative oil profiles versus time (from Carvajal et al. 2015). Soft blue line is water injection only, red line is gas injection only, dotted purple line is traditional WAG injection, and green line is WAGCV injection. *(Taken with permission from EAGE white paper 2214-4609.)*

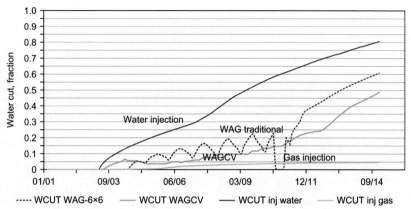

Fig. 7.21 WAGCV simulation: results of water cut profile versus time (from Carvajal et al. 2015). *(Taken with permission from EAGE white paper 2214-4609.)*

2.0 million bbl (13%) more than the traditional WAG process and 5.5 million (44%) more than gas or water injection. Fig. 7.21 shows water cut versus time for the four processes.

This new approach uses an advanced optimization technique that proactively simulates (using 3D numerical simulation) where and when to inject the required slug. The results demonstrate that when using this kind of EOR injection, oil recovery can be enhanced by 5% compared with the traditional WAG process and 15% compared with classical water injection. The simulation showed that water cut is reduced significantly and GOR is kept very low, helping to extend the life of the reservoir production.

7.8.2 Thermal Monitoring

In heavy oils, high-viscosity reservoir in Canada, Shell patented the idea of using ICVs with a steam-assisted gravity drainage (SAGD) process. SAGD (Butler and Stephens, 1981) has been implemented since the 1990s, improving the oil-recovery factor in the area of steam chamber generation compared with traditional continuous steam flooding injection. The main problem with SAGD is the difficulty in controlling the fingering of steam chambers, which causes an abrupt steam breakthrough to the producer wells. To control the steam chamber growth, Clark et al. (2010) have used four ICVs spaced at 200 m each in the steam injector wells. They also used full EOR closed-loop reservoir management tools, which incorporated seismic thermal response, fiber optic, and full-equipped wells with both pressure and temperature gauges to monitor in real time the deviation

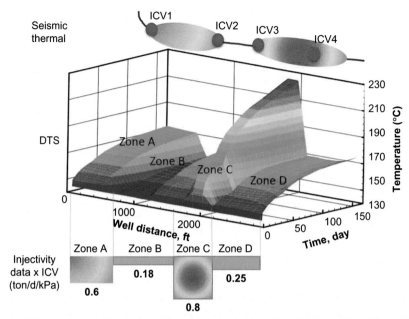

Fig. 7.22 Application of thermal ICVs, seismic thermal profiles, DTS fiber optic in a SAGD heavy oil well. *(Modified from Regtien, J.M., 2010. Extending the Smart Fields Concept to Enhanced Oil Recovery. SPE-136034-MS. https://doi.org/10.2118/136034-MS.)*

in stream quality, pressure, and volume. Their objective was to demonstrate the technical feasibility of using ICVs in a high-temperature environment. They demonstrated that ICVs help to improve the steam injection conformance to heat up the cooler zone (bypassed by steam) and control the steam chamber growth homogeneously across the horizontal section.

Fig. 7.22 shows a combination of the modern reservoir monitoring techniques to control steam chambers. The seismic thermal response can be observed in the top figure, the DTS data profile for both injector and producer in the middle, and the injectivity data through the ICVs in the bottom. It clearly shows that DTS and injectivity data (from ICV) confirmed conformance problems of excessive steam volume in zone C which have been suggested early by seismic thermal response.

7.8.3 Automated EOR/Chemical Process

Injecting chemical products such as alkaline, polymers, surfactants, and low-salinity solutions into the reservoir are processes that improve the oil recovery reserves extraction by altering reservoir wettability, oil displacement mechanism, and residual oil saturation. However, chemical processes have had technical and economic challenges to widespread utilization. By implementing automation (DOF) principles, chemical EOR should gain efficacy as described in this section. By definition, the alkaline is a base (i.e., soap) that easily dissolves in water, and the solution base has a pH greater than 7.0. In the oil industry, the alkaline solution has been used to reduce the interfacial tension (IFT) of the remaining oil in situ, altering the original rock wettability and generating a reduction in the residual oil saturation (S_{orw}) after the primary water injection. Polymers are a large chain of molecules (synthetic or natural) that if injected with the water injection process can increase both the viscosity and the density of the water. The polymers are used to reduce the water mobility (ρ_w/μ_w) in the formation and therefore generate a uniform sweep efficiency displacing oil. Surfactants (also called micelle) are organic compounds that reduce the IFT between different fluids (oil-water); surfactants are used as emulsions or foam agents to absorb the oil phase and generate miscible displacement (one fluid) between oil and water. These three types of chemical injection can also be mixed, designated as ASP (alkaline, surfactant, and polymers injection). The sequence of injection, slug size, total volume, chemical concentration, brine concentration, and injection rate depend on reservoir property distributions across the field.

The chemical injection strongly depends on dominant forces governing the reservoir; these are viscous, gravity, and capillary forces. The reduction of interfacial forces and residual oil saturation can be explained using the capillary number expression.

$$N_c = \frac{Q_w \times \mu_w}{A \times \sigma \times \cos\theta} \qquad (7.8)$$

where μ is the displacing fluid, Q_w is the displacing Darcy flow rate, θ is the contact angle between oil and water, and σ is the IFT between the displacing fluid and the displaced fluid (oil).

In the situation where gravity has a significant component between forces, the potential gradient between the displacing and displaced fluids generates gravitational forces dominated by the density differences (oil-water) but countered by capillary effect; the bond number or buoyancy factor can be expressed as.

$$N_B = \frac{k \times g \times \Delta\rho}{\sigma} \tag{7.9}$$

where k is reservoir permeability, g is the relative gravity of the Earth, $\Delta\rho$ is the density difference between the displacing fluid and the displaced fluid (oil), and σ is the IFT between the displacing fluid and the displaced fluid (oil).

The ASP injection provides a series of challenges: mainly to optimize the fluid injection at reservoir scale. Fadili et al. (2009) described that the main issues related to injecting ASP in any forms can be: (a) to control the solution viscosity over time, (b) non-Newtonian behavior, (c) matrix permeability reduction due to the absorption of the polymers to the formation, (d) capillary de-saturation effect due to fast decrease in IFT, (e) significant losses of chemical component due to the adsorption of the rock, particularly clays and carbonates, and (f) fast chemical degradation.

For DOF, Fadili et al. (2009) suggest that automated EOR projects should be designed to monitor the chemical within the reservoir and be able to adapt quickly to the injection and production schedule automatically. This could be the key element in optimizing the EOR operations. Early water breakthrough of the ASP injection means poor oil-sweep efficiency. Therefore, it requires a series of down-hole completion equipment, such as ICVs, ICDs, and packers to control the water influx, or permanent monitoring tools, such as 4D seismic to monitor the waterfront. However in an ASP project, it is more important to measure in real time all properties that are measured in the lab, but tested at the field.

Chemical properties of the water injection—for example, chlorides, ions, pH, viscosity, density (specific water) properties—can be sampled at the injection stream and compared with production flowback after the separation system, including the emulsion meters, which can be used to measure the quantity of emulsion in oil after the surfactant injection. Tank levels for alkaline, surfactant, polymers, and salt products can be monitored and surveyed daily and generate alarms and alerts in the case of troubleshooting.

Fig. 7.23 shows a prototype of an ASP injection. Tanks, pumps, cyclones, filters, and turbines at the EOR treatments can be automated and set up with PLC or RTU units to monitor status, injection rate, pressure, temperature, and equipment performance (power consumption). Signals are sent in real time using WiFi-WiMAX to a SCADA center. The injection system is set up with preliminary values as follows:

- pH set up to 9.5 (alkaline injection to reduce the IFT in oil-wet reservoir).

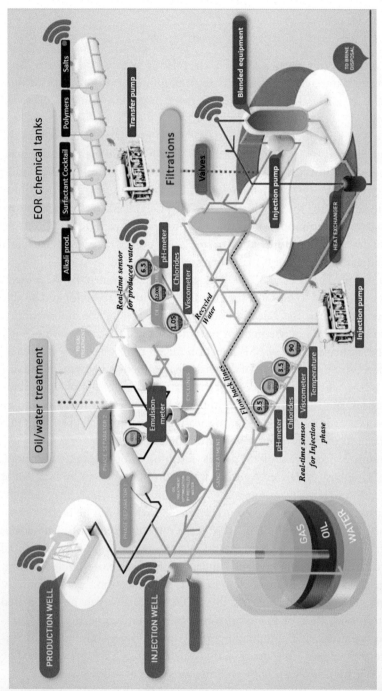

Fig. 7.23 Examples of automated ASP injection showing real-time sensors to monitor pH, chlorides, viscosity, temperature, and pressure at different levels of injection and production chains.

- Chlorides of 10,000 ppm, the required water salinity to generate an optimum balance for the surfactant injection and reduce its absorption level to the rock.
- Water viscosity of 3.5 cP, the required viscosity for the polymer cocktail to displace the oil with an appropriate mobility ratio less than 10.0 $(k_{ro}/\mu_o \times k_{rw}/\mu_w)$.
- Surface temperature of around 90°F, the maximum temperature to reduce the polymer degradation.

Fig. 7.23 also shows chemical sensors setup at the oil and water treatment in the separation system. Before the water breakthrough, it is expected that produced water from the well formation has original chemical levels such as pH of approximately 6.0, chlorides of approximately 25,000–35,000 ppm, viscosity of around 1.0 cP, and no emulsions or micro-emulsion formed in oils. After breakthrough, it is essential to monitor and survey the trend and tendency of pH, ions, chlorides, emulsions, and water/oil viscosity through time in producer wells.

Chemical sensors and a 3D numerical model coupled with an optimizer should be integrated to improve the injection and maximize the oil-recovery factor, while reducing ASP costs. A numerical model performs calculations on gravity, viscosity, and capillary forces. The injection can be adapted and controlled by

- Changing the ICV choke size.
- Monitoring the chemical propagation into the reservoir using preexisting or observed wells.
- Using traditional production behavior, that is, water cut%, GOR, and chemical tracers for ASP and by comparing with chemical sensors shown in Fig. 7.23.

The optimization process should aim to optimize the required injected pore volume (PV) of ASP agents (generally PV can reach a value of 1–2 at reservoir condition) to maximize the oil production (the barrels of oil per $/pound of ASP are incremental) by changing

- Optimum size of the chemical slug.
- Rate of injection.
- Mixed, sequential, or alternated alkaline, surfactant, polymers, or all.

REFERENCES

Ajayi, A., Konopczynski, M., 2003. A Dynamic Optimization Technique for Simulation of Multiple-Zone Intelligent Well Systems in a Reservoir Development. SPE-83963-MS. https://doi.org/10.2118/83963-MS.

Ajayi, A., Konopczynski, M., 2008. In: Intelligent wells technology-fundamental and practical applications. Training material.A Course Introduced at the SPE-ATCE, Denver, CO, 20-21 September 2008.

Birchenko, V.M., 2010. Analytical Modelling of Wells With Inflow Control Devices. Doctoral Thesis Dissertation. Institute of Petroleum Engineering, Heriot-Watt University, Edinburgh, Scotland, UK, 2010.

Butler, R.M., Stephens, D.J., 1981. The Gravity Drainage of Steam-heated Heavy Oil to Parallel Horizontal Wells. PETSOC-81-02-07. https://doi.org/10.2118/81-02-07.

Carvajal, G., Torres, M., 2014. In: Coupling AICD to a wellbore and numerical model simulation.Presentation About AICD Progress. Presented at the Novel Techniques for Reservoir Management, SPE, Frontier of Technology—Reservoir Technology Forum Held in New Mexico, USA, 4–9 November.

Carvajal, G., Saldierna, N., Querales, M., Thornton, K., Loiza, J., 2013a. Coupling Reservoir and Well Completion Simulators for Intelligent Multi-Lateral Wells: Part 1. SPE-164815-MS. https://doi.org/10.2118/164815-MS.

Carvajal, G., Wang, F., Lopez, C., Cullick, S., Al-Jasmi, A., Goel, H.K., 2013b. Optimizing the Waterflooding Performance of a Carbonate Reservoir With Internal Control Valves. SPE-164814-MS. https://doi.org/10.2118/164814-MS.

Carvajal, G., Konopczynski, M., Chacon, A., Mogollon, J., 2015. In: Smart water-alternating-gas EOR using downhole control valve (WAG-CV): concepts, tools and numerical optimization. EAGE-2214.Presented at the 77th EAGE Conference and Exhibition, Amsterdam, The Netherland, 01 June 2015.

Clark, H., Ascanio, F., Van Kruijsdijk, C., Chavarria, J., Zatka, M., Yahyai, A., Shaw, A., Bedry, M., 2010. Method to Improve Thermal EOR Performance Using Intelligent Well Technology: Orion SAGD Field Trial. SPE-137133-MS. https://doi.org/10.2118/137133-MS.

Coats, B.K., Fleming, G.C., Watts, J.W., Rame, M., Shiralkar, G., 2004. A Generalized Wellbore and Surface Facility Model, Fully Coupled to a Reservoir Simulator. SPE-87913-PA. https://doi.org/10.2118/87913-PA.

Cullick, A.S., Sukkestad, T., 2010. Smart Operations With Intelligent Well Systems. SPE-126246-MS. https://doi.org/10.2118/126246-MS.

Daneshy, A.A., Guo, B., Krasnov, V., Zimin, S., 2010. ICD Design: Revisiting Objectives and Techniques. SPE-133234-MS. https://doi.org/10.2118/133234-MS.

Elmsallati, S., Davies, D., 2005. Automatic Optimisation of Infinite Variable Control Valves. IPTC-10319-MS. https://doi.org/10.2523/IPTC-10319-MS.

Essen, G.V., Rezapour, A., Van den Hof, P., Jansen, J.D., 2010. In: Integrated dynamic optimization and control in reservoir engineering using locally identified linear models. IEEE 978-1-4244-7744-9.Presented at 49th IEEE Conference on Decision and Control, Atlanta, GA, 15-17 December 2010.

Fadili, A., Kristensen, M., Jaime Moreno, J., 2009. Smart Integrated Chemical EOR Simulation. IPTC-13762-MS. https://doi.org/10.2523/IPTC-13762-MS.

Greci, S., Least, B., Tayloe, G., 2014. Testing Results: Erosion Testing Confirms the Reliability of the Fluidic Diode Type Autonomous Inflow Control Device. SPE-172077-MS. https://doi.org/10.2118/172077-MS.

Henriksen, K.H., Gule, E.I., Augustine, J.R., 2006. Case Study: The Application of Inflow Control Devices in the Troll Oil Field. SPE-100308-MS. https://doi.org/10.2118/100308-MS.

Jansen, J.D., 2001. Smart Wells. Delft University of Technology, Department of Applied Earth Sciences, Section Petroleum Engineering. Rijswijk, The Netherlands. 2001.

Kharghoria, A., Zhang, F., Li, R., Jalali, Y., 2002. Application of Distributed Electrical Measurements and Inflow Control in Horizontal Wells Under Bottom-Water Drive. SPE-78275-MS. https://doi.org/10.2118/78275-MS.

Konopczynski, M., Ajayi, A., 2004. Design of Intelligent Well Downhole Valves for Adjustable Flow Control. SPE-90664-MS. https://doi.org/10.2118/90664-MS.
Konopczynski, M., Ajayi, A., 2008. Reservoir Surveillance, Production Optimisation and Smart Workflows for Smart Fields–A Guide for Developing and Implementing Reservoir Management Philosophies and Operating Guidelines in Next Generation Fields. SPE-112244-MS. https://doi.org/10.2118/112244-MS.
Least, B., Greci, S., Conway, R., Ufford, A., 2012. Autonomous ICD Single Phase Testing. SPE-160165-MS. https://doi.org/10.2118/160165-MS.
Least, B., Greci, S., Wilemon, A., Ufford, A., 2013. Autonomous ICD Range 3B Single-Phase Testing. SPE-166285-MS. https://doi.org/10.2118/166285-MS.
Nikolaou, M., Cullick, S., Saputelli, L., 2006. Production Optimization: A Moving-Horizon Approach. SPE-99358-MS. https://doi.org/10.2118/99358-MS.
Perkins, T.K., 1993. Critical and Subcritical Flow of Multiphase Mixtures Through Chokes. SPE-20633-PA. https://doi.org/10.2118/20633-PA.
Regtien, J.M., 2010. Extending the Smart Fields Concept to Enhanced Oil Recovery. SPE-136034-MS. https://doi.org/10.2118/136034-MS.
Saputelli, L., Ramirez, K., Chegin, J., Cullick, S., 2009. Waterflood Recovery Optimization Using Intelligent Wells and Decision Analysis. SPE-120509-MS. https://doi.org/10.2118/120509-MS.
Shiralkar, G., Fleming, G.C., Watts, J.W., Wong, T.W., Coat, B.K., Mossbarger, R., et al., 2005. Development and Field Application of a High Performance, Unstructured Simulator with Parallel Capability. SPE-93080-MS. https://doi.org/10.2118/93080-MS.
Twerda, A., Nennie, E., Alberts, G., Leemhuis, A., Widdershoven, C., 2011. To ICD or Not to ICD? A Techno-Economic Analysis of Different Control Strategies Applied to a Thin Oil Rim Field Case. SPE-140970-MS. https://doi.org/10.2118/140970-MS.
Zhu, D., Hill, J., 2008. Understanding the Roles of Inflow Control Devices (ICDs) in Optimizing Horizontal-Well Performance. Distinguished Lecture Program at SPE, a Web Event, SPE Foundation. Available from: https://webevents.spe.org/products/the-role-of-inflow-control-devices-in-optimizing-horizontal-well-performance-spe-distinguished-lecture.

FURTHER READING

Birchenko, V.M., Muradov, K.M. and Davies, D.R., 2010. Reduction of the horizontal well's heel-toe effect with inflow control devices. https://doi.org/10.1016/j.petrol.2010.11.013.

Transitioning to Effective DOF Enabled by Collaboration and Management of Change

Contents

The previous chapters have presented a suite of technologies related to the digital oil field (DOF) and pointed to examples of industry investment in sensor, communication, and automation technologies. However, technology is only part of the requirements for effective DOF that is only a quarter of the whole of DOF components (see lower left vertex in Fig. 8.1). Baken (2016a,b) correctly points out that the rate of return on DOF investments in technology is limited to less than 25%; 75% of expected value can only be achieved through the implementation of DOF with respect to work processes, competency, and role transition and how people work and collaborate using technology that is the upper and right vertices in Fig. 8.1. Ultimately, the three components defined by these triangles must be fully integrated through collaborative work processes. Section 1.5 introduced the collaboration and work processes as critical components of DOF. This chapter presents details on challenges in delivering high value through

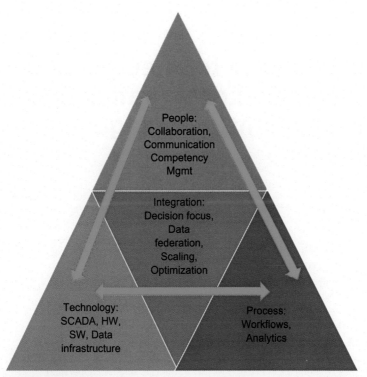

Fig. 8.1 Value from DOF requires integration of people, process, and technology through collaboration.

high-performing teams in DOF, and then discusses the value chain of components and characteristics for the success in change management and collaboration. These components are: (1) the physical space, the collaborative work environment (CWE) and associated mobile technology; (2) team composition and roles; (3) interskill and team collaboration through change management and work processes; and (4) competency development and sustainability.

8.1 TRANSITION TO DOF

The companies have approached a transition to DOF in a variety of ways. We have found that a systematic approach is crucial to make DOF work so that it adds significant value. In referring back to the survey of companies listed in Section 1.8, we see a spectrum of approaches to plan and implement DOF, which include: (1) top-down: a corporate dictate endorsed by executive management and often driven by a centralized R&D,

Table 8.1 Approaches for a New Corporate DOF Initiative

Category	Attributes	Pros	Challenges
Top-down	Executive management or central technology or IT plan a DOF strategy for the company and dictate its take-up.	• Management endorsement • Budget allocated • Good tech support from centralized functions	• Bureaucracy delays implementation • Lots of reporting "up the chain" • Central tech or IT solutions do not meet the practical needs of assets
Bottom-up	Individual asset or business unit decides to implement DOF components.	• Fast-track projects • Focus on exact needs and get quick value	• Budget • Disconnect with other assets; hinders full integration and consistency for systems
Hybrid	Business unit-driven strategy and supported by technology and IT functions; exec management may endorse.	• Projects driven by needs of asset and yet have endorsement, support and budget from exec and central tech functions • Ensure corporate standards rule so multiple assets can coordinate/integrate systems	• Staying focused • Integration across assets

engineering and/or IT departments; (2) bottom-up: a business unit or asset take initiative and demonstrate value on its production and this success gains adherents throughout the corporation; and (3) hybrid: executive management endorses DOF as a concept and provides some "seed funding" but allows asset units to take initiative on pilot projects and implementation. Whatever approach is taken in any given company, DOF implementation requires careful planning. A hybrid (flexible) approach is often most effective and illustrated with an example below. Table 8.1 summarizes these approaches and some of the pros and challenges of each.

8.1.1 Planning a DOF Implementation

The flow chart (Fig. 8.2) provides a planning process that we have used in a several companies with business units to facilitate the planning and

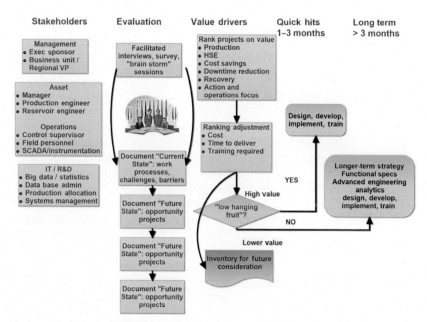

Fig. 8.2 Facilitated planning for a DOF implementation.

implementation of DOF projects. The process follows a hybrid approach discussed in the previous section and is in part based on Eldred et al. (2015). It relies primarily on a business unit or asset level with executive sponsorship and central IT or R&D support. A facilitator works with all the *stakeholders* to identify the *key value drivers* for transition of *"current state"* processes to those of a DOF-enabled *"future state."* In contrast to a "top-down" approach, the stakeholders, that is, management, production, reservoir, and operations, first define their current state through a survey, white-board sessions, and interviews. The interviews are conducted as listening sessions to garner information on each stakeholder's needs with typical questions such as "what keeps you up at night?," "walk us through a typical day at work," "what issues, if solved, would make you more successful?," etc. This process is followed by a focus on what new technology, work processes, roles, and integration could deliver in value either incrementally or as a step change from the "current state," that is, what a "future state" could deliver.

A ranking process of "future state" processes leads to "low-hanging fruit" projects that have potential to deliver very high value in a short period with relatively little cost. By doing a ranking, these projects develop buy-in and a common purpose and stimulate the management and asset team to ultimately support long-term, more complex, and potentially more costly components of DOF.

Fig. 8.3 shows an example from an asset which used the process described and identified its current state as one of essentially supervisory control and data acquisition (SCADA) and alarm monitoring. They developed a vision and then a plan for a progression of DOF phases. The first phase, "low hanging fruit," included automating alarm management and then a phased progression to ever more complex utilization of the sensor data, management by exception, and ultimately integrated operations with proactive actions.

The value of DOF is enabled by collaboration, ultimately how personnel work. Collaboration has been documented as a "must" to add value (Gilman and Kuhn, 2012). Gilman and Kuhn (2012) states that collaboration is not a stand-alone solution but a key enabler. They present a collaboration maturity model. The following is an example of how maturity was planned in stages by one operator.

A large independent oil company planned its transition for unconventional field operations as a series of phases along the lines of the above schematic. A phased approach enables an asset to make incremental additions to the value of the asset, learning by doing, gain credibility in the organization, and justify incremental expenditures on things such as a collaboration room (Section 8.2), well instrumentation, well controls, and role transition and training (see below). The details of each of the four phases are described below:

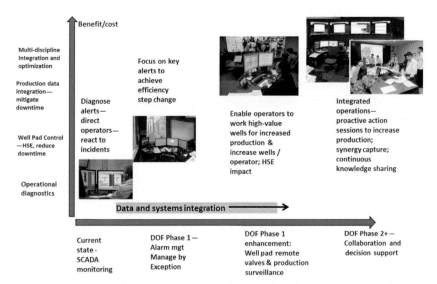

Fig. 8.3 DOF progression plan, from current state to collaborative integrated operations. *(Photos with permission from SPE and ConocoPhillips.)*

Phase 1: Enable pump by exception
- data integrity and cleanup
- production surveillance operations with basic dashboard and rationalized reporting
- production surveillance engineering dashboard for well surveillance
- alarm rationalization
- basic intelligent alarms and data validation and conditioning
- basic well pad automation evaluation
- collaboration evaluation

Phase 2: Enable integrated production operations
- advanced integrated production operations (IPO) dashboards
- automated well review by exception and by well priority and make well review on demand
- alerts for wells producing or not
- automated downtime code analysis and volumes
- intelligent alarms (leaks, predictive maintenance)
- in-line/real-time data validation, cleanup, and conditioning (SCADA data, PVT, and BTU calculations)
- field notes/text notes system upgrade and rationalization
- ticket system for field interventions
- basic well pad regulatory control (choke, gas lift)

Phase 3: Enable collaborative optimization
- collaboration protocols established
- virtual metering in place
- advanced, automated well reviews
- target setting (subsurface collaboration)
- automated gas lift optimization/opportunity notifications with predictive analytics suggestions and data-driven models
- automated rod pump optimization/opportunity notifications with predictive analytics suggestions and data driven models
- flare management
- facilities automation
- advanced human–machine interface (HMI) design

Phase 4: Enable (or at least work toward) "closed-loop" asset optimization including production, operations, and reservoir
- closed-loop optimal regulation and capacity management
- facility scheduling and balancing
- daily route optimization
- integrated field planning and delivery
- production downtime sheltering

Table 8.2 An Example of KPIs for a DOF Implementation With an Operations Center

Category	Establish Baseline Benchmark	Metric: Targets and Actuals for Future Period
Downtime	Deferment; mean time to re-establish production	Reduced deferment: reduced mean time to reestablish production
HSE	Operator miles traveled per week	Reduced miles and vehicle maintenance
HSE	Engineer miles traveled per week	Reduced miles and vehicle maintenance
HSE	Flare issues	Flare compliance and time on flare
Operating Expense	$/BOE	$/BOE vs. baseline
Operations	Number wells per staff	Number of wells per staff KPI
Production	Oil production rate	
Production	Oil production rate	Enhancement from optimizing gas lift
Production	Oil production rate	Enhancement from reduced offset frac deferments
Production	Rod pump caused deferments	Reduced mechanical and workover downtime deferments
Well reviews	Number per week	By exception and opportunity: real time, number per day

8.1.2 Key Performance Metrics for DOF Implementation

A critical aspect of planning is to establish performance metrics and how they will be tracked. Each implementation requires a set of metrics for each phase. Table 8.2 illustrates an example (which could be for Phase 1 for a project implementation as described above) which provides a baseline and targets to be tracked for the production and DOF implementation.

8.2 COLLABORATIVE WORK ENVIRONMENT

This section describes key aspects of a CWE, which include the physical space, the value of collaborative work processes, and the role of mobility. It also provides some examples of collaboration and mobility in practice.

8.2.1 Physical Space

Traditional operation centers with multiple large monitors were initially designed to monitor a single engineering focus (e.g., drilling or facility operations) and were staffed by the appropriate discipline experts. The DOF is evolving into new physical spaces called collaborative work environments

(CWEs) or decision support centers (DSC), which are designed for collaboration from multiple disciplines using fit-for-purpose workflow visualization that integrate across an operational value chain. These spaces can be categorized (depending on their business operations) as meeting rooms, situation rooms, real-time operation centers (RTOC), and CWE (which is typically considered the most advanced). The visualization requirements for monitors and screens to display in real time, and multiple sources of raw and processed data were presented in Chapter 1.

For the KwIDF project, Kuwait Oil Company had two large collaboration rooms (Al-Jasmi et al., 2013). The paper has a schematic diagram of the extensive center in Ahmadi, which has two large collaboration areas and a strategy team room, which can be opened or divided by motorized retractable walls. Each area has multiple large flat panel touch-enabled screens for displaying real-time data from multiple data sources, for example, artificial lift, production allocation, pressures, temperatures, flow rates, power units, etc. Multiple desks have networked, motorized adjustable displays. KwIDF for the North Kuwait asset is intended to monitor, diagnose, and operate from these centers and be staffed by multidiscipline teams (Al-Jasmi et al., 2013). As a somewhat different approach, Shell established fit-for-purpose centers throughout the company as summarized in Table 8.3 (Van den Berg et al., 2014; Van den Berg et al., 2015a,b). Van den Berg et al. (2015a,b) describe the fit-for-purpose centers summarized in Table 8.3 and illustrated in Fig. 8.4.

Goodwin et al. (2010) have reported that BP's advanced collaborative environment (ACE) programs required significant expenditure but provide significant benefits (Section 8.2.2). ACEs were established around the world for BP assets and included onshore operations support centers for offshore operations, including centers for the Gulf of Mexico, UK North sea sector, Trinidad and Tobago, and Azerbaijan. A common thread was that the centers had the communication, videoconferencing, and collaborative work areas so that multidiscipline teams could make decisions in real time on a variety of producing operations.

From reading descriptions of the KOC's KwIDF, BP, and Shell collaborative centers, one might conclude that implementing a CWE means large expenditures, even in millions of dollars for dedicated large centers. However, in our experiences working with small and mid-sized independent operators and the fact that equipment costs continue to decrease, a CWE can be designed at a relatively low cost, depending on the objective

Table 8.3 Summary of Shell's Suite of Collaboration Centers (Van den Berg et al., 2014)

Center	Primary Function	Decision or Output
Real Time Surveillance Centers	Remote real-time monitoring of wells across assets. Automated analytics of issues	Generate and evaluate alerts that require actions by other personnel
Equipment Surveillance	Environments	24/7 monitoring of equipment, e.g., rotating equipment evaluated by experts
Alert onsite staff to equipment issues		
Integrated Operations Centers	Integration of core field team with other teams, e.g., maintenance, procurement, subsurface, etc.	Align and coordinate activities for cross-function decisions
Development CWEs	Integrate production and development activities; collaborate across locations	Optimize development planning

Fig. 8.4 Collaborative work environment work zone and associated decision support meeting room. *(From Van Den Berg, F.G., McCallum, G.A.R., Wallace, S., 2015. Collaborative Working in Shell—Value Achieved, More to Follow. SPE 176787. https://doi.org/10.2118/ 176787-MS used with permission from SPE.)*

and scope of the work to be performed. In our experience with five implementations in three companies for North America assets, a CWE can be achieved at a cost (in the period 2012–16) of $75,000–$300,000 to set up a room with large panel screens and dedicated terminals, servers, and communication equipment. Gilman and Kuhn (2012) point out that significant upfront investment in facilities and infrastructure is not really needed to effectuate collaboration. Costs are lowering and distributed staff can work effectively in addition to being colocated.

8.2.2 Value of Collaborative Work Processes

Chapter 1 discussed the value for various companies for implementing DOF. For collaborative work specifically, there are examples of significant value added. For example, Van den Berg et al. (2014) documents quantifiable benefits for seven assets. Each asset tracks specific KPIs, including tangible benefits such as increased production, reduced deferment, man hours saved, improved HSE, less travel, and intangible benefits such as increased motivation, improved communication, and increased trust. In one asset, production increased overall by 3% and time to bring wells back online from down events improved by 50%. The paper emphasizes that new ways of working enable the real-time technology but this requires effective training and coaching.

Van den Berg et al. (2015b, 2016) have reported that Shell has implemented CWEs in the majority of Shell's assets, covering 55% of its production. Shell deployed these on a global scale and simultaneously provided training. A few examples of high value added outcome are (Van den Berg et al., 2015b): (1) an offshore green field with an onshore control center: an increase in production equivalent to 10 thousand barrels of oil, 1500 engineer hours saved, significant travel and offshore time saved; (2) brownfield: 210 a reduction in deferment equivalent to 210 thousand barrels of oil, field lost production avoided and avoidance of process shutdown, and reduced operation expenses (OPEX). Additional case studies are provided in Van den Berg et al. (2015a, 2016).

Goodwin et al. (2010) have reported on BP's ACEs "… have been shown to generate improvements to BP's business performance." Asset teams "altered ways of working and organizational structures." Benefits reported included: increased production, lower costs, and improved safety. One specific example reported was a 1% operating efficiency improvement of an Azerbaijan asset. In the Na Kika operations center, they reported documented gains of $1 M per month. Key value addition was also associated with the so-called soft benefits such as improvements in working culture, team participation, and collaboration.

Al-Jasmi et al. (2013) present a detailed case study of how collaboration leads to more efficient and effective decisions for an artificial lift optimization decision. In this example, a multidiscipline asset team consisting of surveillance, production, and reservoir engineers and management were able to make a decision from real-time data and advanced analytic models to optimize wells and prevent downtime by interacting and collaborating in the "smart" analytic workflows. The asset team interacted to understand pump performance, fluid characteristics, pressures, and PVT conditions to quickly make the best decision on a well to increase production within safe operating boundaries. A production engineer recommended changing electric submersible pump (ESP) frequency and tubing head pressure (THP) using real-time data and an analytic model to obtain a substantial increase in the production from a well. A reservoir engineer was able to validate the recommendation with respect to reservoir support pressure from nearby water injection and a facility engineer checked the flow assurance. Al-Jasmi et al. (2013) documented how smart workflows saved significant engineering time for effective decisions.

Collaborative working has evolved beyond production and surveillance processes, and now includes new mobility and visualization technologies and supports the drive to reduce costs as described in the next section.

8.2.3 Mobility

Operations are being automated as much as possible to make changes in the field remotely, for example, choke settings, power settings, variable speed drives, etc. However, human intervention is required when mechanical or electrical systems fail. Mobile devices, smart phones and tablets, enable collaboration to go beyond the office and to accompany field personnel wherever they may be. Operations can gain significant value by increasing well uptime by using real-time data to direct field personnel to where they are needed most on facilities and wells, to address downtime events (management by exception)l, and, more so in the future, to act proactively on systems to prevent predicted events. Van den Berg et al. (2015a,b) compared explicitly a "before" and "after" implementation of mobile devices for field personnel. In their study, 10 tasks typically done by a field operator, which included multiple manual actions on information, were reduced to just two operations using mobile communications.

Figs. 8.5 and 8.6 illustrate (before and after) a similar process that was implemented in an unconventional field by several operators in 2015 and 2016. Before management by exception and mobile devices were fully integrated, a field operator (pumper, electrician, chemical specialist, workover supervisor, etc.) had to check into a field office each morning to review a list of wells for particular intervention that day, for example, rod pulls, ESP restart, plunger reset. The policy was to visit all wells on a route in a day to check on any other condition or situation (Fig. 8.5). During the day, the

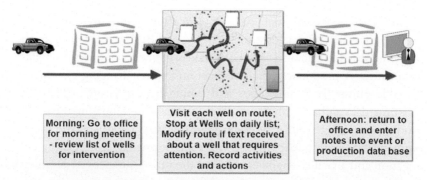

| Morning: Go to office for morning meeting - review list of wells for intervention | Visit each well on route; Stop at Wells on daily list; Modify route if text received about a well that requires attention. Record activities and actions | Afternoon: return to office and enter notes into event or production data base |

Fig. 8.5 Communication process before DOF: field personnel must go to the office to review systems, visit wells to determine issues, and return to office to input event data.

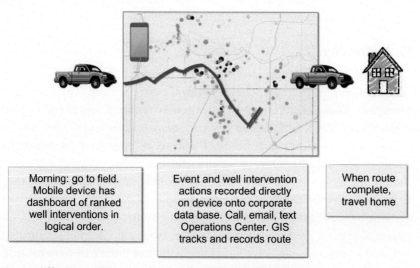

| Morning: go to field. Mobile device has dashboard of ranked well interventions in logical order. | Event and well intervention actions recorded directly on device onto corporate data base. Call, email, text Operations Center. GIS tracks and records route | When route complete, travel home |

Fig. 8.6 Effective mobile communication in enhancing collaboration and action in field, process with DOF implementation: mobile device dashboard directs field personnel to wells requiring attention and event data can be recorded directly.

operations center would text the field personnel with any new alerts or alarms on a well requiring them to make unscheduled stops, which could disrupt the route. Also, a pumper had to visit the well initially to determine if a different technician (such as an electrician, chemical expert, mechanic, HSE expert, or production technician) should be contacted. At the end of the day, the field person would check in at the office to submit any reports and to enter event items onto the company's production or well database in the office.

With DOF field surveillance and management by exception and priority, the personnel can focus on the wells that require the most attention and those that are impacting the production the most (Fig. 8.6). Alerts are rationalized and then ranked by value of intervention (i.e., the lost production), taking into account intervention and workover costs. The alert and alarm dashboard(s) are visible on a mobile device so that the appropriate (based on skill and proximity) personnel can be directed to the most valuable wells, in a logical order and depending on the severity of the situation. Any intervention activity can be logged directly into a production or well database from the mobile device. The location of the vehicle is tracked via GPS into a GIS map visible to the operators and logged into a data base. Analytics can be performed on the travel time, history, and time at well locations for each activity.

Companies get significant value from transition to a DOF-assisted mobile work process, including the following:

- Reduces well downtime. In the first 2 months, one field experienced an almost 30% improvement.
- Reduces operator interaction with lower value wells and increase time on higher potential wells.
- Increases safety through significant reduction in vehicle miles and on-road time (often in rural, hazardous areas), one business unit experienced almost 25% decrease in "vehicle miles-road time" when transitioning to a "by-exception" approach.
- Enables more proactive well maintenance decisions.
- Enables more productivity for operations center staff who can focus on the most important alerts.

8.2.4 Examples: Collaboration and Mobility in Practice

Permian Basin artificial lift. A field in the Permian Basin was experiencing unacceptable downtime and intervention costs for their vertical wells on rod pumps. The wells had pump controllers and real-time data (dynacard, strokes per minute, power diagnostics, etc.) were stored on a contractor's server, but the data was not available online within the company. Generally, data from previous day's event reports (e.g., gyro runs, chemical treatments, tubing and rod scan inspections, etc.) were on separate data folders and physical reports. Each morning the production engineers would manually download all the data onto their desktops and then port the data into spreadsheets so they could analyze report data to prepare for the 8 a.m. "pull meeting."

Supervisor, engineers, and pumpers would convene in the main office and field office to review wells requiring intervention and plan the day's routes for field personnel to then work on the routes.

The company automated the communication and data analytics from the contractor's server and moved all the data folders onto a dashboard available on all computers in the company and on mobile devices (electronic tablets). All stakeholders were able to see the wells identified for the day's action and reasons for the downtime or events. This new way of working meant that the supervisor and engineer could collaborate and issue ticket instructions to pumpers directly over mobile phones and tablets. After the system was in place, the asset reported a 30% improvement in efficiency for downtime and intervention metrics. Production engineers and supervisors were able to spend more time on higher value activity, and pumpers were focused on the well requirements as described above (Fig. 8.6).

Rocky Mountain oil production. An oil field in the central Rocky Mountains had unacceptable downtime in winter months from freezing of the low-pressure gas lines at gathering junctions, which caused the wells to shut-in upstream. Each morning in the field office, field personnel reviewed the electronic field measurements (EFM) of temperature and pressure in flow lines and compressor stations. However, the data were not integrated and could not be analyzed together, so it was not efficient to pinpoint exact locations of bottlenecks. This situation means that to locate bottlenecks, field personnel had to drive to a number of potential sites over a large geographic area using gravel roads in the mountains (like the scenario in Fig. 8.5). Under these rugged and slow conditions, it often took 2 days to get a well back online.

The solution was a new dashboard (Fig. 8.7) with automated data integration and analytics of the EFM to identify bottlenecks in real time. Wells were color coded by time-dependent status based on variations in flow rate, static pressure, and pressure change analytics, including rate of change in a sensor-measured value. Data validation and conditioning were applied on the real-time data (as described in Chapter 3). The dashboard also displayed compressor pressures and flow, power efficiency, and capacity and sent real-time notifications to supervisors and field personnel to address the wells the same day (similarly to Fig. 8.6).

San Joaquin Valley cyclic steam production of heavy oil. Eldred et al. (2015) describes a San Joaquin Valley California cyclic steam project; hundreds of wells were being cycled on steam on varying cycles of injection, soak, and production. In addition, the shallow production and induced fractures

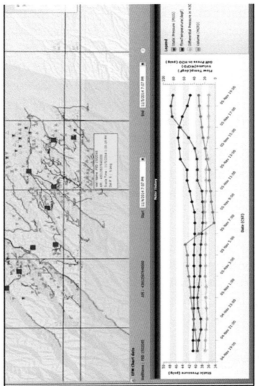

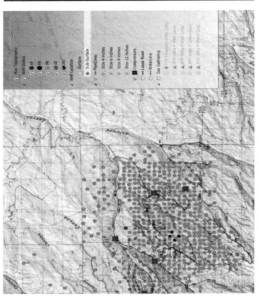

Fig. 8.7 Rocky Mountain surveillance dashboard for pressure analytics on bottlenecks in remote mountain wells.

led to a risk of surface events and well failures. The management of the system required real-time production/injection pressure and temperature data analytics for each well and analysis of surface tilt meter surveys. A key challenge was that much of this analysis was done manually: each morning (7 days per week), a production engineer was tasked with analyzing pdf and spreadsheet reports and comparing it with the production and injection data used to issue daily instructions on wells to direct field personnel.

The solution was an integrated system that displayed all the tilt meter, fracture diagnostics, production, and injection data in one dashboard (Eldred et al., 2015, Fig. 8.8). All stakeholders could then see all the data in a unified environment. Production engineers, reservoir engineers, supervisors, and field personnel collaborated on decisions on cycles; the decisions were not dependent on a single engineer's view at 6 a.m., but benefited from collaborative decisions from the stakeholders using real-time data throughout the day (see next section). As in the other example, field personnel could focus on the wells that required attention and do it more frequently (each half-day). The asset reduced risk of well failure and maintained production more consistently.

8.3 MANAGEMENT OF CHANGE

8.3.1 Collaboration in Practice: "A Day in the Life" of a DOF Operation

Section 1.5 of Chapter 1 discussed how traditionally, disciplines involved in the reservoir management value chain worked in discipline silos, with multiple manual data handoffs, use of different systems, and inefficient communication. Fig. 1.10 shows how a traditional organization of discipline silos can be transformed into collaborative teams. Al-Jasmi et al. (2013) describe how a work team in a CWE makes a decision in real time to change a well operation (artificial lift pump settings) to increase oil rate.

With DOF, data is produced continuously and in real time. DOF systems deliver continuous automated analytics of data and a continuous need for all the asset team members, production, and operations, to collaborate and interact on decisions that drive value. For example, consider an operation with more than 1000 wells on artificial lift with treatment facilities for gas, oil, and water that must be produced with minimum downtime, maximum hydrocarbon production, and zero HSE incidents. Fig. 8.9 shows an example of activity that continued on an operation throughout a 24-h period.

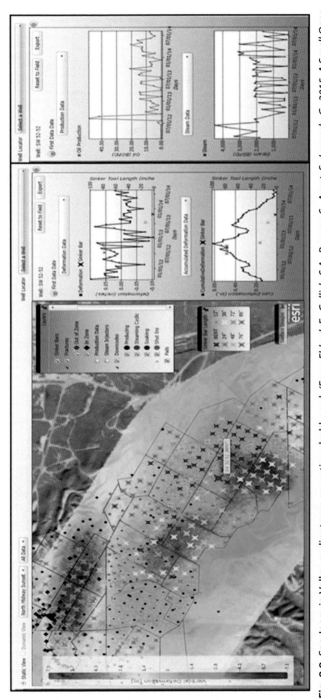

Fig. 8.8 San Joaquin Valley cyclic steam operation dashboard. (From Eldred, F., Cullick, S.A., Purwar, S., Arcot, S., Lenzsch, C., 2015. A Small Operator's Implementation of a Digital Oil-Field Initiative. SPE-173404-MS. https://doi.org/10.2118/173404-MS used with permission from SPE.)

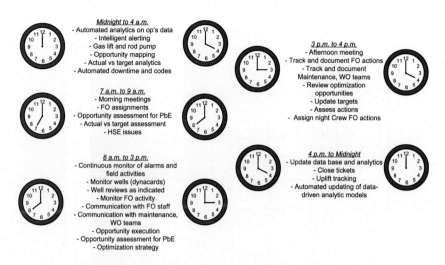

Fig. 8.9 Example of activities in a 24-h DOF operational setting.

Figs. 8.10 and 8.11 show the combination of automated and manual activities that can occur among the disciplines that interact with the raw data and analytics, and that communicate with each other from and to a CWE, offices, and the field (mobile) to make impactful decisions throughout the day.

8.3.2 Change Management: High-Performance Teams

Business and industry literature consistently emphasizes the importance of "people" over "technology." That is, without a change in management process focusing on team formation and dynamics with role clarity that has been planned and implemented, technology investments (e.g., for a collaboration center and other technical investments for well control, hardware/software, etc.) will most likely not achieve the expected value. Fig. 8.12 shows a model based on Tuckman (2016) and Tuckman and Jensen (1997) on the effectiveness of team's progress over time, assuming good leadership and management support. Teams go through stages of formation, objective setting, establishing relationships and role responsibilities, and ultimately to high effectiveness as a team, which Tuckman refers to as forming, storming, norming, performing, and adjourning, respectively. Over time, identification as an individual declines in relative importance as individuals identify more with the team, which results in improved team performance.

Lyden and Zernigue (2014) present an extensive case study from Chevron for building an effective team for a DOF solution project, in part based on the Tuckman model (Tuckman and Jensen, 1997). They summarized the

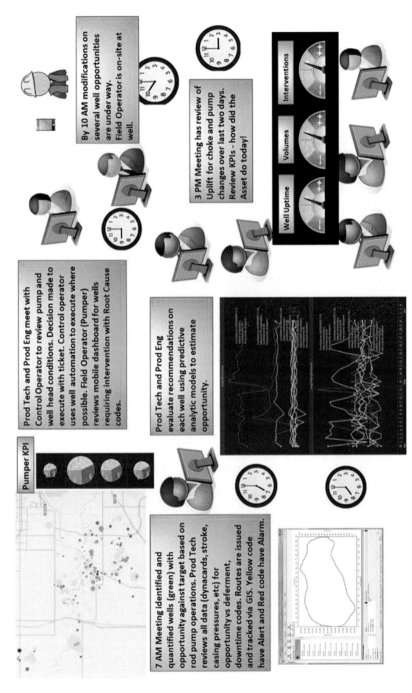

Fig. 8.10 Collaboration activities address downtime alarms, gas lift, and rod pump opportunities.

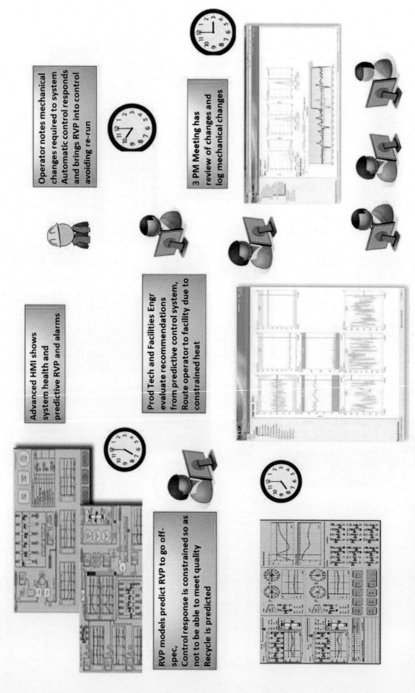

Fig. 8.11 Facilities optimization team addresses off-spec product specifications (e.g., pressure RVP) and uptime.

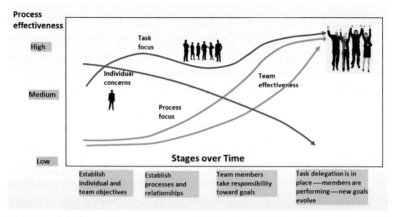

Fig. 8.12 Model of how teams progress over time to high effectiveness through step-wise facilitation. *(Based on Tuckman, B., 2016, MindTools. www.mindtools.com.)*

effort with "transportable lessons" for success, which include: (1) to share accountability, team members must have clear alignment and full commitment to organizational or project goals; (2) team's behavioral expectations must be clearly defined and reinforced by management; (3) a team's decision-making authority and process requirements must be clearly defined; (4) teams must know rules for keeping each other informed and how to interact across distance; (5) the skills required for each specific team must be clearly defined and developed ("monitor and coach"); and (6) celebrate success and assimilate new members effectively.

8.3.2.1 Competency Development

For many decades, the oil and gas industry has provided staff development opportunities which have included formal courses, external and internal, mentoring, diverse job changes, apprentice situations, etc. Industry has in place formal processes for learning and competency evaluation. However, intelligent DOF puts a premium on these learning opportunities and competency development for engineering quality, data analytics, and collaborative work processes. This section discusses examples of learning management (training), competency management, and their synergy, which is often thought of as knowledge management (KM).

A global leader in oil and gas training, development, and competency management defines competency as "as set of defined and observable skills, knowledge, abilities, and behaviors required to perform a specific job" (IHRDC, 2014). The first part of this definition, "skills" and "knowledge" relates to specific job skills that are acquired through on-the-job learning, mentoring, and formalized training.

Virtually for all companies skill development is a high priority for their staff. Skill development, including training courses, has traditionally had a discipline focus. DOF requires additional multi- and cross-discipline training for multidiscipline skills and collaboration behaviors. In addition, some new DOF roles often require staff to experience a "role transition"—which are changes in a traditional role (e.g., production engineer) in a DOF environment. This section describes a training template, a competency model, and an example of a role transition.

The delivery of training ranges from in-house courses to external providers such as professional societies (SPE, AAPG, SPWLA) and commercial companies, such as IHRDC and PetroSkills.

We recommend that each company, through the annual review process, take a comprehensive approach to planning skill development that includes both traditional discipline-specific skills and additional requirements for DOF. Fig. 8.13 shows an example of competency list that one management planned to have for one North American asset's operation center personnel.

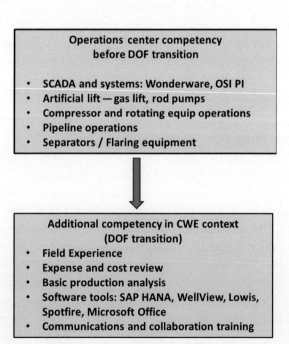

Fig. 8.13 Example of transition of competency requirements for an operations supervisor for DOF/CWE deployment.

Training roadmap: Provide managers and technical professionals a structured road map for training to meet Corporate needs

Workflow hierarchy	Course level definition
Company needs (first level)	Basic:
↓	• Designed for both technical and business oriented professionals who are either new to the upstream oil and gas industry or experienced in one part, but could benefit from a wider point of view.
Professional discipline (second level)	Foundation:
↓	• Designed for individuals who have applied the skill taught in the basic course or those with at least some awareness of basic engineering and operations.
Course discipline (third level)	Intermediate:
↓	• Designed for individuals who require practical familiarity and fundamental insight of the course material. The user desires to put foundational skills into action with an emphasis on active learning and solving problems at a higher level.
Vendor (fourth level)	Specialized:
↓	• Designed for individuals who have completed an intermediate course covering a similar topic or comparable training. courses are focused on a discipline specialty catered to experienced individuals that desire a mastering of skill who are going into or working within a specialized technical area.
Courses (fifth level)	

Fig. 8.14 Example HR training dashboard and roadmap for staff to research and select appropriate training courses.

Before implementing DOF and CWE, the asset team worked with and needed to understand and operate systems related to SCADA, artificial lift wells, and the processing plant equipment. The implementation of DOF required additional skills and knowledge as shown in the figure.

To support and facilitate training, one company provides a learning management system (LMS) dashboard with links to corporate-approved training courses and resources. Fig. 8.14 illustrates this dashboard configured to the specific needs of the company as identified annually by management. The example focuses on reservoir, production, completions, facilities, and some field operations. The dashboard enables staff to navigate a hierarchy of courses to quickly link to many resource sites to find training best suited for them and to communicate with their management. This dashboard has direct links to more than 20 external sites for specific courses and a hierarchy for competency levels to guide the staff in course selection and outcomes. The first page of the dashboard (Fig. 8.14) has all the main discipline areas that are linked to groups of courses and competency levels. Figs. 8.15 and 8.16 show the course options if a user selects artificial lift, followed by selection of Production Operations and Production Engineering Basics.

8.3.2.2 Competency Management

The LMS supports the individual staff member and the direct or immediate supervisor in planning an employee's formal training. An important

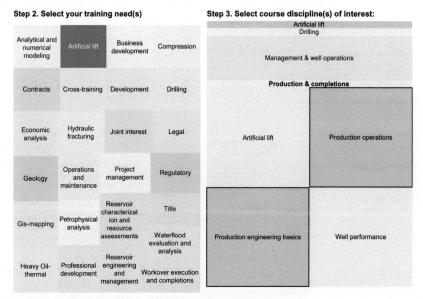

Fig. 8.15 Artificial lift options highlighted for training courses.

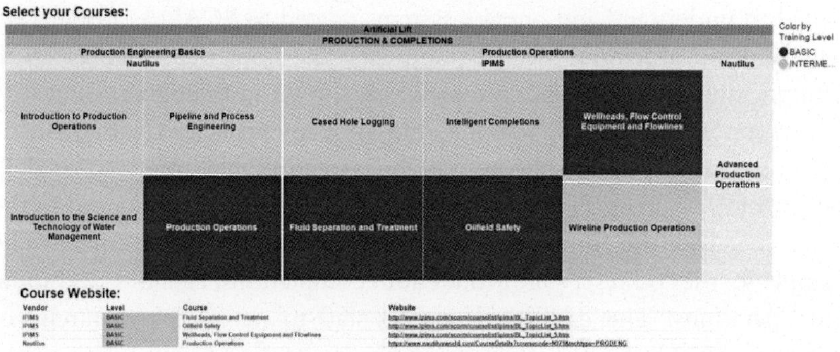

Fig. 8.16 Artificial lift specific links to training courses.

complement to the LMS is the competency management system (CMS), which is a corporate system that maps technical disciplines and experience levels with the competencies that are appropriate for that level and ties those to experience and any formal training (LMS). As defined above, competency is "as set of defined and observable skills, knowledge, abilities, and behaviors required to perform a specific job" (IHRDC, 2014). CMS is a system to manage each company's competency model. CMS as software dashboards are available commercially (e.g., IHRDC and PetroSkills), professional societies (SPE and SEG), and also some companies have their own.

A CMS has a series of competency levels with required skills for each discipline. A CMS might be more specific for the specific needs of a given business unit or geographic area, for example, unconventional onshore, deep water offshore, mature field, improved recovery, etc. For DOF, companies have introduced new competency requirements related to communication and collaboration. Different entities may use somewhat different nomenclature to define level requirements for their competency model. Table 8.4 presents a set of attributes associated with one company's model with three levels: foundational, proficient, and mastery. Other companies often have a variation with four levels so as to have granularity between intermediate application and skilled application. A CMS should have reports for management to track their organization's progress in training and experience goals and within the context of the levels expected, and for individual staff to plan and track progress.

Table 8.4 Definitions for Competency Levels

Foundational (awareness and basic)
- May not have had opportunity to use skill on a project
- Has basic understanding of topic area
- Has demonstrated skill at minimum level
- Able to provide basic assistance
- Has close supervision

Proficient (intermediate or skilled application)
- Able to describe task and process
- Can perform routine activities with limited supervision
- Recognizes when to elevate/consult on complex situations to more experienced resources
- Understands impacts on related systems and processes
- Can identify abnormal situations
- May supervise others in direct tasks
- Able to participate in peer reviews

Role model (mastery)
- Thorough understanding of topic or process
- Regarded as expert in field
- Able to troubleshoot and train others
- May manage a team to execute integrated work, that is, can coordinate work activities
- Establishes trusted advisor status with management, partners, vendors
- Plans integrated activities for teams
- Resource for peer reviews

The SPE has partnered with IHRDC to offer a competency management tool (CMT), which has an individualized dashboard to identify learning gaps, a plan of action, and progress toward the goals. The dashboard is a valuable tool for both individual staff members and management. The CMT has 41 competency areas and 8 job functions: Production Engineering and Operations; Project and Facilities Engineering; Business Development (Operating Company); Supply Chain; Health, Safety and Environment; Business Development (Service Companies); Subsurface; and Engineering (Entry Level).

8.3.2.3 Knowledge Management

KM refers to systems and processes that store, organize, and retrieve knowledge to improve the efficiency of collaboration for learning teams. KM tools help companies capture and manage knowledge and integrate knowledge and experience within the operational systems.

Gilman and Kuhn (2012) describe a KM process wherein the system captures correlations between actions and their results from multiple asset teams. The system can correlate similar situations to help users "explore previous events, issues and solutions" to relate to a current situation in the field.

David (2016) describes a comprehensive KM system that enables employees to leverage collective knowledge and experience of experts. The systems allow employees to discuss the strategy, methodology, and use cases in establishing a collaborative KM foundation. The system features that enable this collaborative foundation include an integrated knowledge base, KM workflows, virtual community of practices, data analytics, and collaboration tools to embed knowledge as part of routine operational workflows.

8.3.2.4 Team Synergy, Behaviors, and Role Transition

Chapter 1, Section 1.6 discussed that, although helping build successful and effective teams is vital for DOF, it's just a part of the DOF change management challenge. For maximum effectiveness of DOF teams (i.e., gains in operational efficiency and output), teams working in DOF environments need to be liberated, authorized, stimulated, and pampered. This type of change involves many other stakeholders around the core business and has consequences for the organization (the way work is structured and organized) and for the way they are managed (leadership and culture), which is fundamentally different from existing traditional ways based on siloed separation of disciplines. High-performance teams do not just happen by management making assignments (Fig. 8.12). Teams go through a process of training, facilitation, and experience that transforms the way they work and a resultant high achievement. Fig. 8.12 presented a model of how teams progress from Tuckman and Jensen (1997). Gilman and Kuhn (2012) discuss how team dynamics and synergy affect DOF implementation and value creation.

Transition to DOF from a current state often involves changing the way specific decisions are made and that may require a transition in staff roles. Staff training and competency have been discussed above, and working collaboratively in a team is critical to success. People often have to learn a new way of working (Gilman and Kuhn, 2012; Goodwin et al., 2010; Van den Berg et al., 2016), which requires management support, coaching, and training as appropriate. Even with these resources, in some cases, some people cannot (or do not) transition well to this new way of working.

An example is the North American unconventional field operation. Current state: The company had three field offices, each with a control room and staff roles approximating those in Fig. 8.17; the control room served as a "coordinator" of daily activity for the production technicians and field operators and a had a dotted line reporting relationship with the production engineer. To transition to a DOF ("future state") the Asset planned to move to a program of intervention by exception and opportunity identification by implementing capabilities such as intelligent alarms, automation of well reviews, downtime alerts, automated event recording including lockouts, and basic well controls, which required changes in roles for operators, both in field and control room, and for production tech and engineer.

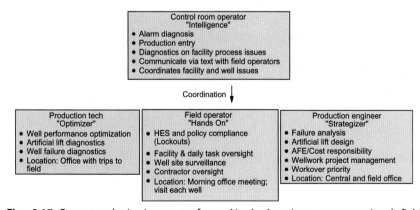

Fig. 8.17 Current roles/assignments for a North American unconventional field operation.

Planning for DOF technology thus led to new work roles and tasks. First, the multiple control rooms were consolidated with data automation, which enabled fewer manual diagnoses and provided a single point of contact. The control room operator and field operator roles were enhanced with a rotation plan and new promotional opportunities. New training and skill reviews were planned. The control room also could function as a collaboration center because the production tech and production engineer would be colocated.

Many reports were automated that previously had manual steps. These included well events, down volumes, actual versus targets, quality assurance, well downtime alerts, action logs, well status lists, flare alerts, lockout histories, rig moves, and more. Intelligent alarming enabled the control room supervisor and operators to focus field activity on the most value-adding activity. This approach requires a change in the way of working, that is, to be more proactive in diagnosing issues and looking for optimization opportunities, which requires more in-depth knowledge of facility equipment, artificial lift, etc. Production techs were taking on more to optimize artificial lift tactically so that the production engineers could become more strategic and work field economics.

8.4 CONCLUSION

Many companies have implemented collaborative working environments, from comprehensive, state-of-the-art centralized rooms to smaller and less expensive venues. Each approach is a move toward a new way of working that can help companies realize significant improvements in safety, efficiency, and operational and economic performance. However, to capture and maximize the value from investment in DOF technology, collaboration and management of change are essential. Companies are implementing new ways of working, helping to train and transition staff from traditional roles to modern roles where people are more proactive, engaged and collaborating in the day-to-day work of the DOF.

REFERENCES

Al-Jasmi, A., Goel, H.K., Rodriguez, J.A., Carvajal, G.A., Velasquez, G., Scott, M., et al., 2013. Maximizing the value of real-time operations for diagnostic and optimization at the right time. In: SPE-163696-MS, https://doi.org/10.2118/163696-MS.

Baken, A., 2016a. Intelligent Oilfields will not deliver full value in the next years. https://www.linkedin.com/pulse/digitalization-take-over-collaboration-andr%C3%A9-baken?trk=prof-post.

Baken, A., 2016b. The toughest Digital Oilfield job of all: creating a high performance team. https://www.linkedin.com/pulse/toughest-digital-oilfield-job-all-creating-high-team-andr%C3%A9-baken?trk=prof-post.

David, R.M., 2016. Accelerating the Learning Curve of Next Generation Exploration and Production Workforce through Smart Data and Knowledge Environment. Society of Petroleum Engineers. SPE-181594-MS, https://doi.org/10.2118/181594-MS.

Eldred, F., Cullick, S.A., Purwar, S., Arcot, S., Lenzsch, C., 2015. A Small Operator's Implementation of a Digital Oil-Field Initiative. Society of Petroleum Engineers. SPE-173404-MS, https://doi.org/10.2118/173404-MS.

Gilman, H., Kuhn, M., 2012. Collaboration as a Cornerstone of Intelligent Energy: How Collaboration is Fundamental to Accelerate and Support IE Success. Society of Petroleum Engineers. SPE-150016-MS, https://doi.org/10.2118/150016-MS.

Goodwin, S., Ford, A., Griffiths, P., Moore-Cernoch, K., Williams, P., 2010. Advanced Collaborative Environments—The Growth of a Capability Transformation Programme. Society of Petroleum Engineers. SPE-128650-MS, https://doi.org/10.2118/128650-MS.

International Human Resources Development Corporation (IHRDC), 2014. Linking Competencies with an Integrated Talent Management Philosophy. https://www.ihrdc.com/pdfs/IHRDC-Linking-Competencies-with-Integrated-Talent-Management-Philosophy.pdf.

Lyden, J., Zernigue, A., 2014. Building Effective Teams to Enable Project Success. Society of Petroleum Engineers. SPE-167843-MS, https://doi.org/10.2118/167843-MS.

Tuckman, B., 2016. MindTools. www.mindtools.com.

Tuckman, B., Jensen, M., 1997. Stages of small-group development revisited. Group Org. Manag. 2 (4), 419–427.

Van den Berg, F.G., McCallum, G.A.R., Graves, M., Heath, E., 2014. Collaborative Work Environments Deployed at Global Scale. Society of Petroleum Engineers. SPE-167871-MS, https://doi.org/10.2118/167871-MS.

Van den Berg, F.G., Chevis, M., Wynot, J., Amos, M., Alharthy, S., Awobadejo, M., 2015a. Collaboration and Surveillance Enabling Optimal Asset Performance. Society of Petroleum Engineers. SPE 173430, https://doi.org/10.2118/173430-MS.

Van den Berg, F.G., McCallum, G.A.R., Wallace, S., 2015b. Collaborative Working in Shell—Value Achieved, More to Follow. Society of Petroleum Engineers. SPE 176787, https://doi.org/10.2118/176787-MS.

Van den Berg, F.G., Weiss, C., van Dijk, J., et al., 2016. In: Collaborative working—harvesting value and evolving with new technologies.Presented at the SPE Intelligent Energy International Conference and Exhibition, 6–8 September, Aberdeen, Scotland, UK. SPE-181083-MS, https://doi.org/10.2118/181083-MS.

FURTHER READING

Al-Enezi, B.A., Al-Mufarej, M., Anthony, E.R., et al., 2013. Value Generated Through Automated Workflows Using Digital Oilfield Concepts: Case Study. Society of Petroleum Engineers. SPE-167327-MS, https://doi.org/10.2118/167327-MS.

Alhuthali, A.H., Al-ajmi, F.A., Shamrani, S.S., Abitrabi, A.N., 2012. Maximizing the Value of Intelligent Field: Experience and Prospective. Society of Petroleum Engineers. SPE-150116-MS, https://doi.org/10.2118/150116-MS.

Knowledge Management Tools, www.knowledge-management-tools.net.

Nautilus Energy Training, https://www.nautilusworld.com.

Petroskills Oil and Gas Training, www.petroskills.com.

Society of Professional Engineers Competency Management System, www.spe.org/training/cmt.

CHAPTER NINE

The Future Digital Oil Field

Contents

The oil and gas (O&G) industry is transforming rapidly and digital technologies are a considerable factor in that transformation (Nyquist, 2016; Murray and Neill, 2017). However, the changes coming over the next few years have significant potential for even bigger step changes. McAvey (2017), VP for Oil and Gas at Gartner in an interview for Pipeline Magazine, indicated that momentum is picking up for DOF and stated "…results from Gartner's 2017 global CIO survey show that oil and gas companies expect to increase their spending on digitalisation to 28% of their IT budget in 2018 from 19% in 2016… While companies are still focused on preserving cash flow, the high-impact nature of digitalisation is proving so attractive that companies want to spend money on it…" (Pipeline Staff, 2017). In another example, in a 2017 World Oil article Exxon Mobil's chief computational scientist is quoted as saying, the "oil patch's digital transformation will be comparable to horizontal drilling's tech revolution" (Endress, 2017) (which, of course, was the game-changing technology that enabled the shale gas revolution, among other major industry achievements). Furthermore, the report states that "…he predicts the digitalization of oilfield equipment and operations will continue for the foreseeable future, due to future competitive advantages and untold economic value" (Endress, 2017).

This final chapter highlights a few of the exciting technologies that are in development or are being envisioned for the digital oil field (DOF) of the future.

Chapters 1–8 present DOF technology and processes that today are state of the art and have been documented in the technical literature. It is impossible for anyone to know all that is coming in the near future. So we have enlisted some help from some industry experts to provide their insights for technologies they see for the future and highlight a few technical areas that will impact O&G production. Our panel includes company CEOs, engineers, and scientists in companies ranging from large national and international to midsize independent oil companies and service providers.

We hope these chapters have led you to the conclusion that implementing DOF has the potential to transform O&G companies, to allow them to manage their operations and businesses more efficiently. However, through 2017, it is estimated that less than 25% of all companies (national and international oil companies and service companies) have applied or introduced DOF concepts in many of their internal processes. Only 10% of the total process and workflows are actually automated. We can argue that this low implementation rate is due to lack of experience and a slow approach to making internal process changes.

A number of exploration and production (E&P) companies have established that the main priorities of DOF investment can be classified as follows:
- improving people's safety in high-risk areas and protecting process security;
- cybersecurity, data protection, and data sharing;
- expanded connectivity and communication across the fields, operation centers, well locations, separator batteries, tank farms, office, and terminals;

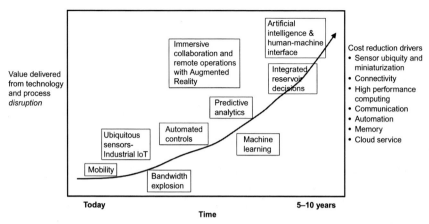

Fig. 9.1 Disruptive technologies that are driving value for DOF implementation.

- unlimited, clear, and clean data collection and sharing across the organization, which is well protected against cyber-attacks;
- easy application of automated processes to reduce cost over existing manual processes;
- improved efficiency in sensors, equipment, devices for real-time control;
- enhanced production levels by reducing production downtime through the adoption of data modeling and software solutions that aid decision-making.

This chapter highlights the following topics for the industry's intelligent DOF of the future. Fig. 9.1 illustrates how these technologies and processes will continue to add value in future DOF implementations as they disrupt current practice and technology:

- ubiquitous sensors, that is, the industrial Internet of things (IIoT),
- data everywhere or Big Data,
- next generation of analytics,
- automation and remote operations,
- knowledge everywhere, knowledge capture, and management,
- integrated reservoir management,
- collaboration, mobility, remote operations, and machine-human interface.

9.1 UBIQUITOUS SENSORS (IIoT)

Experts predict that the number of connected devices is increasing from about 23 billion in 2015 to more than 50 billion in 2020 (Cisco, 2013), and that stored data in 2020 will be 40 zettabytes (40 billion terabytes). For the O&G industry, however, Mahdavi (2017) says that the industry seems uncertain about how to integrate IoT for optimal business impact. He goes on to say that adoption of IoT-enabled analytics in the consumer space, for example, where retailers have so much data and can analyze it so quickly they can predict buyer behavior in real time, has created an expectation by some in O&G that digitalization would also enable real-time proactive decisions. He indicates that the DOF has progressed by enabling collaboration across geographies and through real-time centers, but it has not fulfilled the promise of transformative performance in terms of value delivered. Mahdavi reinforces the report by McAvey (2017) that business transformation is possible by fully integrating the systems "from the sand face to their back-end IT and financial systems." BSquare (2017) says "in fact, IoT technology creates an entirely new asset: information about these crucial elements of their businesses," and goes on to say "through the establishment of comprehensive, data-driven predictive insights, O&G companies can employ sophisticated rules and machine learning to constantly adapt and tune expensive assets in real-time using trend analysis."

In an interview, Anthony McDaniels of Rare Petro Technologies noted that in 2012, a surveillance device that cost USD 10,000 and took 6 months for design, decision, power, communications, and installation can, in 2017, be installed for less than USD 300. The cost for a field data device has gone from $5 per megabyte (MB) to USD 0.03 per MB. Even mature fields with lower production can now justify using surveillance equipment and obtain value with the analytics available. New technology for miniaturization of the power source, wireless communication, and polymer protection enables use of downhole sensors, which could not be considered in the past.

Chris Lenzsch of Dell-EMC spoke about the increase of software-defined sensors (that is edge sensors), more IT integration with the data by users, and increased automation. People will use data that are close to the measurement sources. With edge sensors and data on the cloud, there is an increasing focus on data security. The sensors themselves require built-in security. In the future, DOF systems will be increasingly closed loop with control over decisions and action. People interacting with systems simply to move and manipulate data, for example, between file systems, spreadsheets, reports, data warehouse, etc., will disappear; Lenzsch states, "Today the industry is 70% data manipulation and 30% decisions; in the future, the percentages will be reversed at the least." E&P data will move to the Cloud so that data are "ubiquitous whenever, whatever you want for collaboration and remote operations," Lenzsch predicts. He also referred to how Total just put a robot on a topside in the North Sea—automatically doing what many people were doing before. Lenzsch also indicated that breakthroughs in hardware are leading to $10\times$ greater storage every 3 years—end-memory flash solid state is now terabytes and will be many TB in the near future. Solid state is replacing discs for fast data retrieval and backup, and analytics will be running from a "mother ship" hub that will communicate with multiple devices and will feed models to interacting "edge" sensors.

From the viewpoint of Mohammad Askar of the Advanced Research Center (EXPEC ARC) at Saudi Aramco, the future of O&G operations is through a fully automated self-learning and auto-optimizing system using a variety of sensors, remote sensing devices, and a wildly creative communicative network. Imagine a network of sensors distributed intelligently everywhere in the field, above and underground, around and far from the wellbore, while drilling and post-flowing the wells, during operation, etc. Included with these sensors are another line of remote-sensing devices that sense far in advance (examples include water encroachment before it happens, loss circulation, and all sorts of other operational issues that might arise). Such information (or big data) is acquired in real time, fed to a storage system (cloud drive) that is linked to a high-performance computing system. The simulator will run thousands of realizations in real time, and the results are sent to a smart, self-deciding system, loaded with specific operational scenarios to instruct, for example, choke valves (to control which well contributes to production while keeping water cut low, etc.) and downhole flow control valves (to control or shut down laterals that are either producing too much water or about to) through remote sensing.

The future of sensors is very bright also in the area of production surveillance systems. According to ARC, in Saudi Aramco, one goal is to put the whole field under continuous and real-time deep surveillance, in the most efficient and cost-effective way, through such enabling technologies. This means more data points of different types at various locations, at the surface and downhole. It is only controllable and changeable if one can measure it better in real time. Imagine if one can see what is going on for a whole field or section of the field all at once and in real time!

The new generation of sensors can be enabled by new generations of microprocessor chips. Mims (2017a,b,c) reports on companies like Apple, Nvidia, Intel, and others that are "breaking Moore's Law" with task-specific chips. They move software and applications from the CPU and build the calculations into the chips so that the work can be done many times faster without needing to miniaturize continually the circuits. These new chips enable image processing, image recognition, self-driving cars, virtual reality (VR), and artificial intelligence.

9.1.1 Nanosensors

Nanotechnology refers to miniaturizing technology to a nanoscale so that it can be used in remote places where conventional sensors do not work, like deep in reservoirs. Naturally, O&G operations have plenty of situations where nanotechnologies can be used and will benefit DOF solutions. Fig. 9.2 shows nanosensor film used in high-pressure/temperature pipes. The film nanosensors help visualize the pressure gradient from point to point as they communicate the data to the operations.

Will nanorobots (or nanobots) as machines or designed chemicals be a reality for DOF solutions in the next 5–10 years? In an interview, Askar from Saudi Aramco states, "ResBots are a reality today!" As a research program in EXPEC ARC, the ResBots program started about 8 years ago, with a suite of technologies being developed in-house, to address various existing field challenges, such as: tracing well connectivity, in situ sensing the progression of a waterflood front in real time, in situ determination of the remaining oil saturation, and direct interventions through targeted or on-demand delivery of oil field chemicals to predetermined locations deep in the formation. The idea is to investigate effective ways to access areas far beyond the wellbore region, for sensing and/or intervention purposes.

The industry is always looking to develop technologies to better under-stand/characterize reservoirs or to alter specific subsurface areas (plug, change IFT, or wettability) for better performance. Nano-based solutions in O&G are attractive because: (a) miniaturizing materials to nanoscale gives us the ability, at least physically, to access tiny rock pores (where

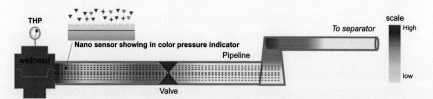

Fig. 9.2 A flow line after the wellhead covered by a film of nanosensors detecting in real time the pressure gradient from the wellhead to the outlet point (separator). The red area means high pressure near the wellhead and valve, whereas the light blue color means low-pressure values.

hydrocarbon exists) and (b) manipulating matter at that scale will enable us to access surfaces by nanomaterials that would not be possible otherwise. In the future, the idea is to develop fully functional and steerable nanomaterials (called nanoagents) that are capable of traversing the rock formation by harvesting their own energy (maybe from the fluid flow or heat of the reservoir), reporting about the reservoir properties, and supporting discovery procedures and oil recovery processes. These nanoagents could store the information in a variety of ways including a detectable change in their optical properties, chemical signatures, etc. Once recovered at the surface, nanoagents can be interpreted by engineers and scientists. For some nanoagent templates, progress has been made beyond lab testing. For example, model templates of nanobased tracers (e.g., A-Dots and advanced tracers) have been deployed in the field in cross-producing wells, an industry-first cross-well field tests.

Autonomous, self-controlling, human-free transportation is becoming the future of civil and cargo transportation. Will the O&G industry see a similar trend toward the autonomous control and automation of technology platforms for future DOF systems? The question was posed to Vasily Demyanov, associate professor at Heriot-Watt University's Institute of Petroleum, who responds that the technology for remote and autonomous sensing has improved considerably in recent years from the breakthrough in scale miniaturization of the monitoring instrumentation. Micro- and nano-scale sensors are now being developed with a current level of technology that reveals new distinct properties of materials, including 2D films. Implementation of such monitoring technology would enrich and make a step change for DOF monitoring. The target here is to collect cheap and abundant information about what is going on in the reservoir and how it responds to interventions.

9.2 DATA EVERYWHERE

Raed Abdalla, CEO of Evinsys, a DOF solution provider, says that the next generation is really going to be in three areas: (1) Data: data will be provided faster, and more real time, but there will be growing need for security because of so many connected devices. We have a real gap right now for data and networks. (2) Data use: there is a drive for analytics and predictive analytics and to store and track the data and predictions. (3) Management of connected devices: there is no holistic system to manage many interconnected devices, that is, monitor and report on the health of devices, quality of data, calibration, etc. He goes on to say that we will see a lot of changes and innovation, such as smart sensors with self-calibration, self-diagnostics, and self-diagnostics on the connectivity with other sensors, and edge devices with analytics and collaboration. An important question is: how much intelligence to put into each edge device with more intelligence which feeds the cloud? There will be a lot of innovation in intelligence in the edge devices for what data are actually transmitted to cloud for storage. Predictive analytics is also coming in a big way for equipment failure, scheduled maintenance, production analysis, and empowering O&G operations to move beyond reaction.

Fig. 9.3 shows a schematic of how the data will be acquired and displayed everywhere. The concept of "data lake" is being introduced to the data ecosystem. Currently, O&G companies mostly have data warehouses and structured and relational data for their master data stores, which have developed over several decades. Transactional industries are transitioning to new architectures for Big Data and IIoT. A data lake stores data, both structured and unstructured, in a form close to its native state, along with metadata characteristics. This approach enables a much more flexible system for data access by the many applications, analytics, machine learning, visualization, integration, etc. that need to access it. Seamless data will be required for the integrated reservoir management discussed later.

Abdalla went on to discuss how the new generation will use much more intuitive technology (software), simple to learn, with no training courses, that is mostly self-guiding workflows. New platforms have "self-protecting" workflows that alert or prevent users from making mistakes. "Apps" will be sources for users to build their own workflows or use existing ones with analytics, for example, predictive maintenance and artificial lift optimization. Currently, there is a gap to define protocols and standards across vendors. IIoT needs a more structured protocol so devices can communicate with each other, rather than one way to the cloud or data storage. Open standard across companies will enable sensors (with "smarts") to communicate among themselves.

9.3 NEXT-GENERATION ANALYTICS

Kunal Dutta-Roy and Senthil Arcot of Technical Toolboxes, Inc. (a global provider of integrated and cloud-based pipeline software and consulting) describe how they see the future of analytics for artificial lift. Pumps and well equipment will be monitored like a car, informing their owner that

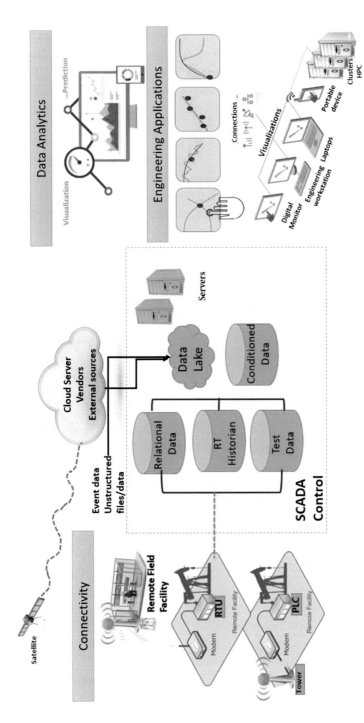

Fig. 9.3 Industrial internet platform connecting real-time data with multiple sensors, and sensors to IIoT to enable analytics, distributing data according to engineering purpose and applications, supporting deployment and real-time visualization at the edge of the cloud, under data security and governance.

oil change or service is needed. There will be no reason for a middleman to make assessments and decisions. Different devices are connected to run a field operation; today, devices have to communicate "up" to a central device or dashboard, but in the future there will be an ecosystem of different end devices communicating with each other by common standards for an open environment. Reaching this vision has some challenges: production operations have not kept up with standards and security might be an issue. The first to market with standards may drive the competition to join in. Maintenance will be recorded in real time and also will guide additional actions. All data will be captured in a knowledge system.

Drones and nanosensors will be used to monitor system leaks. Nanosensors are available now for leak detection; and like all newer technology, prices are declining quickly, nearly 100 times less than a few years ago. Sensors will provide the basis for analytical, data-driven, and modeling solutions. Sensors make modeling (machine learning) more reliable, and will eventually be used in multiple system end points to build models.

Use of a single data lake (as described above) means one integrated data source for an operating asset, which will finally drive movement away from the domain silos that have plagued the industry for decades. Humans will interact and make decisions at a more integrated level. Solutions will be built as services—data as service, analysis as service, visualization as service, alerts as service. In the future, analysis provides guidance and communication to operators to act. Brain power and software are on contract.

There is and will be ubiquitous video at well pads and well sites with remote and automated operations, that pan, zoom, time, recognize abnormalities, record acoustics and then store, analyze, and alert for abnormal issues. Visual diagnostics enable virtual site inspection and will be automated (see Pixel Velocity, 2017) with recording of events and pattern recognition.

Here is an interesting comment from Jim Crompton, Managing Director at Reflections Data Consulting, "Technology vendors are so far advanced versus the current maturity of the upstream O&G operator that the two are struggling to have a constructive conversation ...Digitization of the oilfield (at least many of them) is happening, but that does not mean that effective analytics will naturally follow." The paradigm of the "digital twin," a digital representation of the physical system, is that every asset and its components "learn" using physics plus data analytics from other assets as to best procedures and processes. The knowledge repository is in the "digital twin."

9.4 AUTOMATION AND REMOTE CONTROL

Dr. Demyanov stated that analyzing, inferring, and making decisions based on the new monitoring (see Nanosensors above) would be the next task that would require a step-change level of data analytics implemented into reservoir surveillance workflows. Can we speculate on the transition toward artificial intuition embedded in DOF technology so the O&G digital transformation could result in a self-thinking DOF ecosystem? Demyanov notes that artificial intelligence and cognitive systems have been widely used in mining large data domains and supporting expert judgment by elicitation of vital information patterns from data. Such data-driven methods have been historically implemented in seismic processing before gaining a wider recognition in other O&G application domains. Nowadays, the cognitive approach is intensively implemented in many diagnostic systems that require interaction with human activity/technology, such as medicine, electronics, etc. There is great potential in implementing data-driven cognitive learning-based technology in DOF systems. The present challenges lie with a technological level of storing and mining the right amount of information at the acceptable cost level. Furthermore, the cognitive systems to be developed to mine reservoir operational information should embed experiential learning based on decades of manual operational success and failures. Such technology already exists (e.g., in mechanical manufacturing) where mechanisms are able to capture and learn from the operators' experiences. However, the more complex the mechanisms are, the more elaborate and sophisticated the cognitive learning needs to be, especially given the vast uncertainty in hydrocarbon reservoirs and the variability of many different options and possibilities of events.

Demyanov concludes that there is a great potential in linking the step change in data acquisition in the DOF era with the novel data-driven learning-based workflows—as the second is the solution to extract the added value from the first. It is essential to view the high-level problems of where the value is in the O&G operations and what are the bottlenecks where the value is depreciated for certain reasons. Thus, the efforts need to be aimed at those bottlenecks to explore the opportunities of intelligent data-driven workflows that would be aimed at gaining the qualitative step change in the way the DOF systems are operated. Present-day practices and technology allow monitoring social behavior and reveal certain dependencies and even predict behavior events based on the monitored response of a swarm activity. The understanding and interpretation of such monitoring still requires a lot of effort in tying with the governing dynamics of the complex systems.

In another interview, Anthony McDaniels, President of Rare Petro Inc., addressed remote operations and control. The "technology industrial revolution" is here and is being driven from consumer electronics. Robotics is going to be on site soon and although humans will still be in the field, they will be fewer and fewer.

As hardware and processing become less expensive, there is a "tipping point" to move the 'smarts' (analytics) to the sensors that are on the end devices ("edge"). Even small operators are able to collect data and in the future will have the analytics as well. Therefore, move analytics to the "edge" and provide decisions and guidance to users—users do not have to spend their time managing and analyzing data. Software is automatically upgraded and distributed to devices.

The demographics of the younger generation demands mobile technology for everything. There will be "Uber" for supply chain, well interventions, tank deliveries, etc. With GPS on every truck, pumpers will be tracked and guided for step change in efficiency. Services will be commoditized.

Is there a potential in synergies between the state-of-the-art social/ mobile high technology and O&G optimization, and how could the E&P digital transformation maximize the "bang for the buck"? Pallav Sarma, Chief Scientist at Tachyus (an O&G technology company), notes that, while techniques and underlying principles of data science have been around for decades in various disciplines such as statistics, computer science, machine learning, probability theory, etc., it is only recently that data science as a unifying umbrella has received significant attention and popularity. This popularization is due to an explosion of the quantity of data collected daily by these technology companies, a significant increase in computational resources available, access to cloud computing facilities, and advances in data science algorithms. A recent and highly visible example of the application of data science in technology is the 4-to-1 defeat of the reigning GO champion by Google's DeepMind team. GO is a board game orders of magnitude more computationally complex than chess; therefore, brute-force computational solutions are not viable yet for solving GO. Just recently, it was generally thought that a solution to GO was at least 10 years away. However, DeepMind's approach to GO was made possible by access to huge amounts of training data, access to Google's very large GPU clusters, and significant advances in deep-learning neural networks over the past decade. Furthermore, the O&G industry, while generally lagging in uptake of new technology, has also seen a surge in activity and applications of data science.

However, this is certainly only the beginning of applications of data science to the O&G industry. While such applications to major value drivers like reservoir modeling are currently being explored, there is scope for improvement of almost all workflows that consume data of some sort, and while not being the primary source of value, still have significant impact on the entire value chain. From predictive modeling of oil field maintenance and safety to the applications of block chains for supply chain management, etc., the possibilities are endless!

9.4.1 Wireless Technology

Automation and control will in part be enabled by step changes in communication technology with higher bandwidths. To assure the seven priorities established by the O&G industry (listed at the start of this chapter), wireless technology (Wi-Fi) is a key factor in the future advancements of DOF systems. One such promising technology is Li-Fi. Created by the University of Edinburgh in 2011, Professor Harald Haas tested for the first time a light emitted from diodes (LED) to transmit data to a mobile device. The invention is called Li-Fi, which is defined as a "light-based communication technology that delivers at high speed, bidirectional networked, and mobile transmission in a similar manner as Wi-Fi" (pureLiFi, 2017). Fig. 9.4 illustrates a Li-Fi system. Li-Fi is 10,000 times the frequency spectrum of radio; the data transmission can achieve up to 10 Gbps, whereas Wi-Fi can achieve up to 2 Gbps. In terms of security, Li-Fi cannot penetrate walls; therefore, it is more secure and private than Wi-Fi. However, the Li-Fi cannot penetrate solid objects and sometimes natural sun light, bulbs, and other external light can interfere with the transmission. Another issue is that Li-Fi requires a constant and reliable source of electric power supply. In DOF systems, the great benefit of Li-Fi would be complemented with Wi-Fi. If a reliable source of power is available, a series of electrical LED bulbs can be installed at a well location to transmit data during the night, and during the daylight, traditional Wi-Fi using a solar battery can be used. Fig. 9.3 shows a futuristic location operating during the night; a series of LED lamps illuminates the entire operation of drilling and production sites while transmitting data from the production wellhead location and drilling operations to a data center.

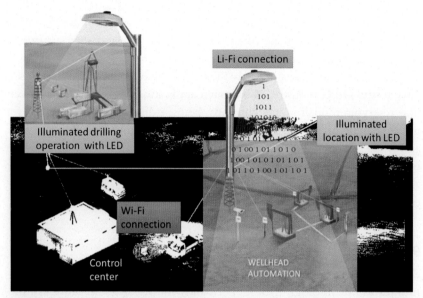

Fig. 9.4 A futuristic production location at night; an LED map illuminates a wellhead production location and drilling operations to transmit real-time data to an operation center.

9.4.2 Drones

Drones are just beginning to be used in O&G fields, but they have the potential to reduce the cost of operations significantly. Today, drones can be equipped with high-definition (HD) cameras, laser-based sensors, Wi-Fi, high-frequency radio frequency, ground-penetrating radar, ultra-sound sensors, and light detectors and optical beams to measure gas emission, read data on gauges, and notify for leaks or for intruders. Drones can be very positive for the industry to generate real-time data for DOF systems helping with the following tasks:

- Visualize entire field operations and safety conditions.
- Surveil remote areas where operators cannot access easily or take too much time to go in truck.
- Surveil areas with a hostile environment where the operational wells are inaccessible due to snow, sand, high temperature, flooding, etc.
- Patrol and investigate for intruders and nonauthorized personnel.
- Track down pipeline leaks and perform methane inspections.
- Map with high-resolution imagery, thermal data, and digital points to generate elevation maps.
- Carry equipment from warehouse to well location to facilitate engineering operations in less time [e.g., supply chain ("Amazon delivery")].

9.5 KNOWLEDGE EVERYWHERE: KNOWLEDGE CAPTURE AND PEOPLE RESOURCES

9.5.1 Capturing Knowledge in New Ways

DOF systems require new ways of working and are leading to new careers. In the past 5 years, a career as a data scientist (with titles such as chief data scientist, chief computational scientist) has become mainstream in both large and small O&G companies and in the service industry. This relatively new role that began in the IT Department a decade ago in many organizations has moved into the operations and business units. Engineers in operations are being tasked with learning and performing data analytics and statistics, "data-driven production optimization," and are increasingly coming from that background and training. This role is necessary because of the use of ubiquitous data as described above and the use of new technology to analyze and use the data. The increasing trend of data science is likely to grow as other conventional roles in the field decline.

9.5.2 Delivering DOF to the Business

Introduction of DOF systems is also bringing innovations to the delivery model. Terms such as data acquisition as a service, analytics as a service, data management as a service, communication as a service, and intervention as a service are entering the industry lexicon. This delivery of services as needed might be called the "Uberization" of many inputs to the decision chain for which DOF is a complex integration of services. Edge sensors (with internal analytics) are being installed and interconnected (IIoT) by service providers on a unit cost basis. Operating companies will not "own," that is, capitalize, the sensors, communications, Big Data storage, etc. Even analytics will become a service, for example, Microsoft's Azure model.

9.6 INTEGRATED RESERVOIR DECISIONS

9.6.1 Big Data and Big Models

To improve production forecast accuracy and better decisions to enhance O&G recovery, Big Data should be integrated in the future into large reservoir models. There is no official definition for big reservoir models, but we can understand big models as those high definition/resolution 3D gridded models that preserve the geological heterogeneity and fluid properties in cells with fine resolution. Today, technology is available to manage massive data, more than 100 TB, and soft-computing hardware and high-performance clusters to speed the CPU time. However, the E&P industry faces many challenges to integrate Big Data with big models, including:

- Enough storage capacity to submit more than 100 realizations/scenarios to the cloud.
- Parallel multiple simulation jobs without increasing cost.
- CPU scalability and acceleration to reduce CPU.
- Budget constraints. Technology for hardware and software is available, with literally thousands of economic options for cloud and cluster environments.
- Decisions about which real-time data should be integrated into big reservoir models. Monthly production data are enough for 3D reservoir modeling. However, the water and gas breakthrough could occur in weeks. The simulator is capable of predicting when fluid breakthrough will happen and what action should be taken to prevent it.

Do we really need to integrate Big Data with big models? In many experiments, we observed potential discoveries and insights that were not observed with upscaled processes. Stochastic analysis is the key to run Big Data-big model to explore the impact on production forecast and oil recovery, especially when uncertainty plays a fundamental role.

Fig. 9.5 is a schematic for the integration of a big reservoir model and Big Data, applying production data to update the model, running many scenarios for production forecasts, generating intuitive diagnostic and analysis of production downtime, extracting data for data analytics, and showing where to drill, complete, and optimize well production performance. It will be one common platform to capture real-time data into the model to generate scenarios rapidly and rank economic decisions; the future platform will provide intuitive workflows without coding or mapping individual properties to connect different software applications. It is envisioned that the platform will generate cognitive diagnostics to rank solutions according to events and well issues. Models will integrate physics-based and data-driven responses.

9.6.2 Optimizing Optimization and the "Closed Loop"

Chapter 6 presents the state of the industry in optimization and introduces the closed-loop concept. The future DOF will harness and benefit fundamentally from advancements in the process optimization technology. According to Pallav Sarma, an integrated monitoring and control approach known as model-based closed-loop optimal control has to be implemented

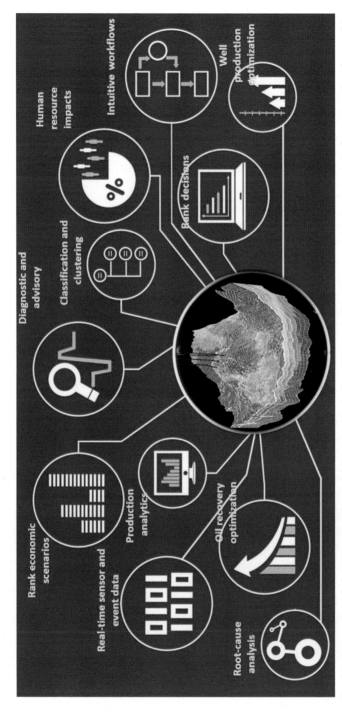

Fig. 9.5 Integration of a big 3D reservoir model (high resolution) and Big Data. Showing many visualization and information plots for an all-in-one integrated reservoir decision.

to extract the maximum benefit from the enhanced monitoring capacity and controllability of smart fields. The goal of such a system is in essence to continually maximize the life cycle value of the oil field by enabling proactive optimization and decision-making, which in turn is enabled by real-time monitoring, continuous model updating, and optimal control of the oil field. Sarma also notes that the key components required to enable a closed-loop optimal control system are as follows: a continuous data acquisition and integration system, a set of forward models relating the control variables to performance indicators or objectives, ability to update these models with the latest data, ability to optimize these models across multiple objectives and constraints, and finally a system that enables easy consumption and implementation of decisions recommended by the system.

New sources of real-time data such as fiber optics and permanent downhole sensors have increased the volume of data collected by orders of magnitude. However, much of these data are not used to the fullest extent possible, and almost certainly not for proactive decision-making. For example, pump-off-control and other well-related data can be used for predicting the probability of well failure and thereby allow predictive maintenance. However, maintenance decisions are still typically reactive: wells are fixed after they fail. Such reactive solutions are in general "too little, too late" and can be quite costly.

The second component of the closed loop, namely, the forward model, has almost always been a reservoir simulation model in existing (partial) implementations of the closed loop. However, the computational complexity of simulation models, coupled with the significant time and effort to build these models, makes the closed loop all but impractical to apply in anything close to real time. Although techniques can be applied to alleviate some of these problems, there is no complete solution today and this is undeniably the most critical component and also the one hindering mass adoption of closed-loop control.

The third component of the closed loop is technology to efficiently and accurately update the forward model with the latest data. Techniques here range from well-established deterministic gradient-based methods to more recent stochastic/heuristic methods such as genetic algorithms. An additional aspect of the model updating problem is re-parameterization or dimensionality reduction of the uncertain model parameters. While there is still work to be done, Sarma feels that this is an area where significant progress has been made toward a practical solution that is applicable in a closed loop.

The fourth component of the closed loop is technology to efficiently optimize the decision variables using the updated models that relate these decision variables to multiple objective functions. The techniques used here are similar to those used for model updating, such as gradient-based and stochastic optimization methods. Recent approaches, such as mimetic methods that combine the benefits of both gradient-based and stochastic approaches, can help reach solutions. Further, combined with scalable parallel cloud computing and reduced-order models, the optimization problem has become practically solvable in a closed-loop framework.

Sarma emphasizes that the final and probably most neglected component of the closed-loop system is a system to enable users—from managers to engineers to operators—to easily analyze, collaborate, and implement decisions recommended by the closed-loop system. Oil producers and service companies are notorious for delivering software with very poor and non-intuitive user interfaces, and most such software are static in that they are not connected to live data streams, and are certainly not collaborative, that is to enable multiple disciplines to interact effectively. A software application enabling closed-loop optimization has to be designed from the ground up integrating the best design practices and tools from the software industry to bring this archaic part of the closed-loop system to modern times.

Sarma elaborates that physics-based reservoir simulation and data-driven machine learning offer complementary strengths. An ideal predictive model would combine the speed and flexibility of machine learning with the predictive accuracy of reservoir simulation, at the right scale, so that operators can integrate data in real time to quantitatively optimize reservoir management decisions continuously. To this end, data physics models merge modern data science and the physics of reservoir simulation seamlessly. Data physics models, like machine learning models, require only days to set up and can be run in real time. Additionally, since they include all the same physics as a reservoir simulation, they offer excellent long-term predictive capability even when historical data are sparse, missing, or noisy.

Data physics models are optimized for speed and run orders of magnitude faster. This speed permits repeated comparison and tuning to the underlying history of production and consequently permits statistically quantifiable comparative prediction performance between many alternative scenarios in service of quantitative optimization.

Given this definition of data physics, while this approach is fundamentally different from either traditional machine learning or simulation, on its own and considering only the science behind the technology, it is not sufficiently different to be called a quantum leap in modeling technology. However, since these models are really designed to enable practical implementation of the closed-loop system and provide users with an interface to collaborate and manage all important reservoir management decisions from daily to long-term ones, it can indeed be considered a quantum leap in how it enables near real-time data-driven reservoir management processes not possible to traditional approaches.

9.6.3 High-Performance Computing for the Future DOF

For many industry experts, the advancement in high-performance computing (HPC) and cloud computing are seen as key enablers for the next-generation DOF. The future is now and the technology vendors are aggressively investing in graphic processing unit (GPU)-based reservoir simulation technology (e.g., Stone Ridge Technology (SRT), 2017 with its ECHELON simulator and Rock Flow Dynamics (RFD), 2017 with its tNavigator simulator) with the vision to transform the future DOF. However, as noted in the interview with Vasilii Shelkov, CEO at RFD, in the past few years there has been very little progress in accelerating reservoir simulations, both for shared memory (workstations, laptops) and distributed memory (public and private clouds) systems.

The typical performance improvements expectations from a CPU upgrade have diminished to 5–10%, which quickly became unacceptable and forced the reservoir simulation community to look for new ways to accelerate. Since performance of reservoir simulations is dominated largely by throughput and latency of the memory system, the emergence of general purpose GPU computing and its fast memory started to attract the attention of software developers. As of now, more and more companies show that CPU-GPU and GPU-GPU systems can easily challenge scalar performance of classical CPU-CPU systems. The development of new CPU vector processing tools like AVX512 could make compute-bound parts of reservoir simulations more competitive with GPU, but will not help much with memory-bound calculations. Using general purpose GPU computing for reservoir simulations has already shown potential and with the merge with CPU and memory systems may eventually become next-generation technology to boost reservoir simulation performance by an order of magnitude or more in the next couple of years. In fact, the lack of competition is perceived by Shelkov as the only major hindrance for progress of HPC and cloud computing in the future of the DOF.

According to Shelkov, historically, owing to security concerns, reservoir simulations are typically run on hardware tightly controlled by corporate IT managers. However, as the hardware refresh cycles have shortened, especially with GPUs, the benefits of public clouds have started to attract attention. Using only public clouds, reservoir simulation jobs can be extended to hundreds and even thousands of computing nodes. The mass use of public cloud systems can drastically change the uncertainty analysis landscape.

IBM has just announced the 50-qbit quantum computer as a part of the IBM Q Program. Can quantum computing be seen as a "quantum leap" for the future of reservoir simulation and as such the next-generation DOF? Shelkov believes that this technology seems to be many years away from being useful for reservoir simulations but may eventually become interesting. So far, there have been no attempts to use these systems for reservoir simulations.

9.7 COLLABORATION, MOBILITY, AND MACHINE-HUMAN INTERFACE

9.7.1 Mobility and Collaboration

Mobile applications (apps) on phones and tablets are becoming the norm for field operators who have the same access to data and dashboards as the operations center and engineering staff. Through mobility, decisions are moving closer to the wells. Ailworth (2017) reports on an EOG Resources app being used by rig personnel to change the direction of bits with real-time data and real-time communication with central offices. He quotes Sandeep Bhakhri, EOG's Chief Information Officer, "The apps help employees work at the 'speed of thought'." Other companies are moving apps into the hands of all field personnel, pumpers, instrument technicians, etc.

9.7.2 Virtual and Augmented Reality Enable Immersive Collaboration

VR and augmented reality (AR) are rapidly transitioning from the gaming and entertainment industries to commercial industry and medicine. Fig. 9.6 illustrates remote or "tele" surgery that is not only being tested but is also being implemented in medical centers. The technology requires very high and immediate transmissions of data, voice, video, and analytics. We see VR and AR growing rapidly for O&G in the future.

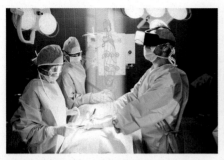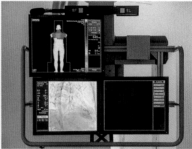

Fig. 9.6 Remote surgery or "tele" surgery: on the left is an operating theater with surgeon using AR for guidance; on the right is a control unit for remote control for angiography.

Another area where VR has a significant impact is in training and education. One page of full text can teach a concept and a process description, one picture can explain the concept, and a 2D image with motion can explain 100 times what a picture shows. However, in a dynamic, immersive environment of 3D and 360-degree views, VR and AR can communicate many times more details of a process. Shell, Chevron, and Exxon-Mobil now have introduced various training programs using VR to reduce training costs and avoid obsolete training materials (documents, power point presentation, etc.). Fig. 9.7A shows AR for a processing plant operation, a digital twin of the operation. Fig. 9.7B–D are from a case study on a Shell Permian operation used for training (Dietz, 2017). The technology is a tablet application that takes 360-degree views of operations in the field, the operations center, processing facilities, wells, etc. The views become a basis for AR dynamic interactions by trainees with all the field equipment and can be done from remote locations. Using VR and AR, the operator can teach at the office without moving, travel, or waiting for instructors. In the near future, it could be shown that with training using VR, operators can significantly reduce operation risks and for engineers, it will be easier to receive more information in a shorter time, with focus on details that affect operational safety.

Another benefit of AR is working closer with operators at the field—taking virtual tours of facilities and well locations without traveling and showing an exact replica of what the operators are viewing in real time and sharing it with people in the office. Managers and engineers become more informed, increase the awareness of everyone on the team, and ultimately make better decisions at the right time. The VR will be a tremendous advantage to reduce misunderstanding and miscommunication across the

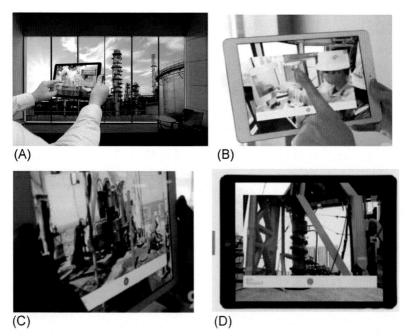

Fig. 9.7 Augmented reality for plant operations and field training. (A) Real-time operations digital twin of a processing plant; (B–D) AR 360-degree view of rig and field operations set for remote training. *(Used with permission of 900LBS of Creative.)*

team. Illogic Digital Creativity Lab (2017) has a virtual processing plant by which a plant operator can interact dynamically with the plant equipment.

Geoscience and subsurface models are representations of the reservoir to quantify the amount of O&G volume in situ and show the main geological features of the reservoir, for example, faults, lithology, water contacts, etc. For nonsubsurface engineers, reservoir models have been considered a black box, difficult to understand. Even for reservoir engineers and geologists, it can be quite difficult to explain what happens in the reservoir when geological heterogeneities play significant roles in oil extraction. Using VR, the entire asset team (all disciplines) can see the unique interpretations (3D-360 degree) of what the reservoir looks like, where the remaining O&G reserves are, and the localized main faults and boundaries of the reservoirs.

As noted by Steve Dietz and Cole Wiser of 900LBS of Creative, we are transitioning from the "digital age" to the "experience age"; that is, the user will be immersed in the content. Conference rooms, collaboration centers, and operations centers will have more scaled 3D visuals with mixed reality for very practical applications. People will be virtually inside of equipment and performing operations on wells and facilities just like remote surgery. Haptic feedback is bringing more real-time cause and effect; peripherals may not be just goggles or glasses but wearable vests that have sensors for pressure, feel, temperature, feeling of elevation, etc. Google Project Tango may enable transition from today's "data anywhere, anytime" to "operational interaction and impact anywhere, anytime." Users will be immersed inside the system with not just visual sense but audio, touch, and "presence." Multiple colleagues will be able to interact and see each other moving in a relative sense, avatar-like, for example, on a rig or facility. Interactions may be tracked for learning with heat maps for the elements engaged.

Chris Lenzsch of Dell-EMC sees that someone in the field will see equipment and operations on their tablet or wearable with a complete history of interventions, visuals from manuals, step-by-step instructions, inside equipment, sensor readings, etc. simultaneously with remote personnel at operations centers.

9.7.3 Human-Machine Interface

Humans and machines that are computing devices are increasingly interacting in ways not previously available. Goldberg (2017) refers to this as "multiplicity," which is computers and multiple humans interacting as a dynamic learning enterprise. Multiplicity is diverse groups of people and machines working together to solve problems; it is not science fiction. A combination of machine learning, the wisdom of crowds, and cloud computing enables continuous learning. Goldberg describes what Google, Facebook, Netflix, etc. do when learning users' preferences and assembling new offerings as a component of this artificial intelligence, multiplicity. Automated driving and robotics are enabled by this machine–human learning. "Collective intelligence with AI enables many of the most sophisticated and effective systems in use today."

Another frontier is telepathy (Mims, 2017a,b,c). Facebook is working on things like headbands that would receive and communicate brain signals to computers and machines. Other companies are working on brain implants.

These technologies taken together yield a "digital twin" that is a digital representation of physical equipment (BSquare, 2017; GE Digital, 2017). The digital twin concept is transitioning from the consumer marketplace to industry. Consumer companies have digital representations of the characteristics of their customers with predictive analytics of what individuals will tend to do or purchase. For industry and the DOF in particular, there can be a digital representation for every component in a value chain; for example, a steam turbine in a plant (GE Digital, 2017) with thousands of parts. The digital representations for pumps, compressors, pipelines, etc. have predictive analytics to alert with respect to maintenance, wear, and production. In the digital twin paradigm, every asset learns from the other assets and the learning is based on physics plus data analytics.

With human-machine interfaces and AI, capabilities to operate oil fields from complete DOF systems are on the horizon.

9.8 SUMMING UP AND LOOKING AHEAD

Section 1.5 introduced the three major components for successful DOF systems: people, technology, and processes (Holland, 2012). We then overlay these high-level components with more detailed components that we call the core of DOF, which has five main areas that must be fully synchronized to implement a successful DOF solution: sensing and control, data management, workflow automation, visualization, and collaboration. The relationship between these components and additional details are shown in Fig. 9.8. In closing, we take a final look at the three main components.

9.8.1 People

In 2008, some early visionaries of the DOF discussed the notion of a *digital petroleum engineer* (DPE) in an article that appeared in the Society of Petroleum Engineers (SPE) *Journal of Petroleum Technology (JPT)* (Mahdavi, 2009). They defined the DPE role at its basic level as someone who "combines IT knowledge with O&G content."

This book certainly supports the notion that skills from multiple domains are vital for successful DOF implementations. As the book, however, has shown, DOF today and in the future, unlike the view in 2008, requires skills much beyond IT, that is, deep discipline engineering (production, reservoir, facilities, and operations), data science, instrumentation and communication, interdisciplinary knowledge, and collaborative interpersonal competencies. We consider this book an "in the trenches" report on DOF design, development, and implementation. It was our intent with the detailed case studies and examples to show that deep knowledge of all these areas is necessary.

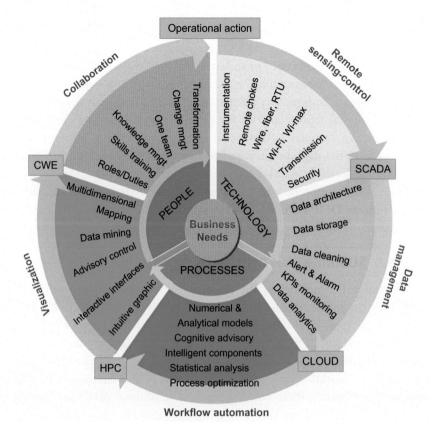

Fig. 9.8 The main components of DOF: a business-centric model where the key areas of people, technology, and process are further organized into five critical components of DOF (shown on the outer ring) and related detailed tasks (in each segment).

Since the technology and how it is used is changing at an ever-faster speed, people working on DOF solutions must be prepared for lifelong learning, through formal training, on the job learning, and independent reading. We encourage you to seek out the many references cited in this book—many of them are papers published by the SPE and available online at the links provided in the reference.

9.8.2 Technology

The same *JPT* article said, "Although we have made progress. It is time to take a deep breath and take the plunge into total immersion, to be like fish in water." Today, this statement is "more true" than when first published. In the past, the O&G industry "applied" technology to specific O&G domains, for example, very early on for reservoir simulation of production analysis. Over the years as technologies progressed and new ones emerged, we have continued to apply them to domain problems with additional efforts to integrate them to support extended workflows.

However, now, with some of the newest technologies we have discussed here—such as Big Data, the cloud, and IIoT—we really must think of developing DOF solutions that immerse discipline-specific solutions in integrated systems and automated workflows. This is an important shift in how we approach development of DOF solutions. Additionally, technologies such as artificial intelligence and machine learning enable us to build truly intelligent solutions that can self-monitor, learn, and intelligently control operations hardware and equipment in the field. These technologies provide an environment for powerful DOF solutions with comprehensive end-to-end workflows, but they also increase the complexity of developing and maintaining these systems. Again, the detailed case studies and examples are intended to show that, while the problems are complex, they are surmountable.

9.8.3 Processes

As technology progresses, it becomes increasingly intertwined with process. Chapters 5, 6, and 8 discuss how technology is used to construct, execute, and manage automated workflows. Successfully creating these workflows not only requires deep knowledge of both current domain practices and workflows, and the technology to correctly implement them—but also another "level" of knowledge to not merely implement current practice but understand both well enough to use the technology to transform current workflows to achieve new levels of efficiency, improve safety, and access more value from an asset.

Naturally, no one person can or will know all of this on their own. Chapter 8 talks about the importance of collaboration and change management. To create successful, comprehensive DOF solutions requires that people work together in new ways, sharing, debating, and deliberating what can be done and how to realize it. Additionally, working in an operation with sophisticated DOF solutions changes how people can work and opens up possibilities for transforming the operation, because much surveillance and response is performed by the DOF system, leaving time for engineers to focus on higher value work.

9.8.4 Use this Book for Your DOF Projects

Our end goal for writing this book is to help individuals see possibilities for their companies and the industry to implement successfully DOF technologies, systems, and solutions so that it can increase value in terms of improved safety, efficiency, and operational and business performance. We hope these real-world case studies and detailed examples and applications help you, companies and individuals, better understand what is involved in implementing DOF solutions—the available approaches, the challenges and how to solve them—and that this book serves as a useful resource for your current and future DOF projects.

REFERENCES

Ailworth, E., 2017. Fracking 2.0: Shale Drillers Pioneer New Ways to Profit in Era of Cheap Oil. Wall Street Journal.

BSquare, 2017. Industrial IoT: From Concept to Business Reality. Webinar. www.bsquare. com.

Cisco, 2013. Embracing The Internet of Everything. http://www.cisco.com/c/dam/en_us/ about/ac79/docs/innov/IoE_Economy.pdf.

Dietz, S., 2017. Shell Case Study. 900LBS of Creative. https://900lbs.com/portfolio/ item-3/.

Endress, A., 2017. Exxon Mobil's Halsey: Oil Patch's Digital Transformation Will Be Comparable to Horizontal Drilling's Tech Revolution. http://www.worldoil.com/ blog/2017/05/15/exxon-mobil-s-halsey-oil-patch-s-digital-transformation-will-be-comparable-to-horizontal-drilling-s-tech-revolution.

GE Digital, 2017. Digital Twin at Work: The Technology that's Changing Industry. White Paper. https://www.ge.com/digital/blog/digital-twin-work-technology-changing-industry.

Goldberg, K., 2017. The Robot-Human Alliance. Wall Street Journal. https://www.wsj. com/.

Holland, D., 2012. Exploiting the Digital Oilfield: 15 Requirements for Business Value, Corporate Edition. Xlibris Corporation LLC, Bloomington, IN.

Illogic Digital Creativity Lab, 2017. VR Star. Available from: http://illogic.us/project/vr-star-virtual-reality-platform/.

Mahdavi, M., 2009. The Digital Petroleum Engineer: Carpe Diem!. SPE-1008-0016-JPT. https://doi.org/10.2118/1008-0016-JPT.

Mahdavi, M., 2017. Ushering in a new era of oilfield innovation with the internet of things. J. Pet. Technol. 69 (7), 14–15. https://www.spe.org/en/jpt/jpt-article-detail/? art=3060.

McAvey, R., 2017. What oil and gas CIOs should do now to prepare for a volatile future. Gartner Report, July.

Mims, C., 2017a. A Hardware Update for the Human Brain. Wall Street Journal. https:// www.wsj.com/.

Mims, C., 2017b. How Chip Designers are Breaking Moore's Law. Wall Street Journal. https://www.wsj.com/.

Mims, C., 2017c. How Facebook's Telepathic Texting is Supposed to Work. Wall Street Journal. https://www.wsj.com/.

Murray, P., Neill, M., 2017. Delivering on the Promise of Digital Transformation. World Oil, pp. 81–84. http://www.worldoil.com/magazine/2017/april-2017/features/delivering-on-the-promise-of-digital-transformation.

Nyquist, S., 2016. The Invisible Revolution that is Re-Defining the Oil-and-Gas Industry. LinkedIn article, Big Ideas & Innovation, Oil & Energy. https://www.linkedin.com/pulse/invisible-revolution-re-defining-oil-and-gas-industry-scott-nyquist.

Pipeline Staff, 2017. Pipeline Oil and Gas Magazine, March 26. Available from: https://www.pipelineme.com/features/interviews/2017/03/gartner-says-momentum-picks-up-for-digital-oilfields/.

Pixel Velocity, 2017. www.pixel-velocity.com.

pureLifi, 2017. http://purelifi.com.

Rock Flow Dynamics, 2017. http://rfdyn.com.

Stone Ridge Technology, 2017. http://stoneridgetechnology.com.

FURTHER READING

Baken, A., 2014. The Ultimate Definition of a Digital Oil Field. http://dofas.info/blog/the-ultimate-definition-of-a-digital-oilfield/.

Beggs, D.H., Brill, J.A., 1973. A study of two-phase flow in inclined pipe. J. Pet. Technol. 25 (5), 11. https://doi.org/10.2118/4007-PA.

Beijia, M., Nahal, S., Tran, F., 2016. Future Reality: Virtual, Augmented & Mixed Reality Primer. White PaperBank of America Merrill Lynch, New York.

Bronk, C., 2014. Hacks on Gas: Energy, Cyber Security, and U.S. Defense. Baker Institute for Public Policy. Available from: https://bakerinstitute.org/research/hacks-gas-energy-cybersecurity-and-us-defense/.

Clayton, B., Segal, A., 2013. Addressing Cyber Threats to Oil and Gas Suppliers. Council on Foreign Relations. Available from: www.cfr.org.

De Souza, A., 2017. The Big Issues in Engineering Simulation: HPC Deployment. Benchmark Magazine, NAFEMS. https://www.nafems.org/publications/benchmark/.

The Economist, 2017. Oil Struggles to Enter the Digital Age. https://www.economist.com/news/business/21720338-talk-digital-oil-rig-may-be-bit-premature-oil-struggles-enter-digital-age.

Falcone, G., Hewitt, G., Alimonti, C., 2009. Multiphase Flow Metering, first ed. vol. 54. Elsevier, Amsterdam.

Graham, E., 2014. Multiphase and Wet-Gas Flow Measurement. UK-National Measurement System. NEL Training Course, Aberdeen, UK.

Gray, W., 1978. Vertical Flow Correlation in Gas Well. API Manual 14BM, Houston, TX.

Greenwald, T., 2017. New Intel Technology Bridges Gap Between Speedy Conventional Memory, Longer-Term Storage. Wall Street Journal. https://www.wsj.com/.

Mills, E., 2013. Saudi Oil Firm Says 30,000 Computers Hit by a Virus. CNET. http://news.cnet.com/8301-1009_3-57501066-83/saudi-oil-firm-says-30000-computers-hit-by-virus/.

National Petroleum Council, 2001. Securing Oil and Natural Gas Infrastructures in the New Economy. http://energy.gov/oe/downloads/securing-oil-and-natural-gas-infrastructures-new-economy.

NIST, 2014. Framework for Improving Critical Infrastructure Cybersecurity. National Institute of Standards and Technology. http://www.nist.gov/cyberframework/upload/cybersecurityframework021214final.pdf.

Norwegian Society for Oil and Gas Measurement, The Norwegian Society of Chartered Technical and Scientific Professionals, 2005. Handbook of Multiphase Flow Metering, second ed. The Norwegian Society of Chartered Technical and Scientific Professionals, Oslo, Norway.

Persons, D., 2014. Energy Pipeline: Cyber-Attacks Hit Oil, Gas, Just as Much as Retail. The Tribune. http://www.greeleytribune.com/news/business/10355602-113/cyber-oil-attacks-security.

Rantala, R., 2004. Cybercrime Against Businesses: Bureau of Justice Statistics Technical Report. US Department of Justice, Washington, DC. https://bjs.gov/content/pub/pdf/cb.pdf.

Roberts, J., 2012. Cyber Threats to Energy Security, as Experienced by Saudi Arabia. Platts. http://blogs.platts.com/2012/11/27/virus_threats/.

Upstream Intelligence, 2017. Turning Data into Dollars: Synthesizing Business Architecture and Optimizing Operations in the Age of Digitization. White Paper, London, http://analysis.upstreamintel.com/.

Winkler, R., 2017. Elon Musk Launches Neuralink to Connect Brains With Computers. Wall Street Journal. https://www.wsj.com/.

INDEX

Note: Page numbers followed by *f* indicate figures, and *t* indicate tables.

Printed in the United States
By Bookmasters